Jürgen Valldorf · Wolfgang Gessner (Eds.)

Advanced Microsystems Automotive Applications 2004

Springer-Verlag Berlin Heidelberg GmbH

Jürgen Valldorf · Wolfgang Gessner (Eds.)

Advanced Microsystems for Automotive Applications 2004

With 198 Figures

 Springer

Dr. Jürgen Valldorf
VDI/VDE-Technologiezentrum Informationstechnik GmbH
Rheinstr. 10B
D-14513 Teltow
e-mail: valldorf@vdivde-it.de

Dr. Wolfgang Gessner
VDI/VDE-Technologiezentrum Informationstechnik GmbH
Rheinstr. 10B
D-14513 Teltow
e-mail: gessner@vdivde-it.de

ISBN 978-3-662-31235-3   ISBN 978-3-540-76989-7 (eBook)
DOI 10.1007/978-3-540-76989-7

Library of Congress Cataloging-in-Publication-Data applied for
A catalog record for this book is available from the Library of Congress.
Bibliographic information published by Die Deutsche Bibliothek.
Die Deutsche Bibliothek lists this publication in the Deutsche Nationalbibliographie;
detailed bibliographic data is available in the Internet at http://dnb.ddb.de

Typesetting: Jasmin Mehrgan
Cover design: deblik Berlin
Printed on acid free paper   68/3020/M - 5 4 3 2 1 0

Preface

Over the past years enormous progress has been made in the microsystems area in transforming research results into marketable products. Nearly all economic sectors did benefit from these developments. Automobiles today are inconceivable without microsystems. New and improved functions in the car related to safety, performance, comfort and emission reduction – being in many cases the unique selling propositions of a new automobile product – are in the majority of cases based on microsystems. ABS, break assistance, stability control and further safety features, applications in air condition systems, powertrain and engine management are only some examples for the indispensable role of microsystems in automobiles. Microsystems and their underlying technologies are in many cases even the driving forces in order to satisfy new customers' requirements.

Even more than in the past microsystems are the keys in order to face the enormous challenges to be faced by the automotive industry in the future:

Vehicles have to satisfy higher standards concerning environmental issues: requirements concerning reduced air pollution, energy consumption and sustainability requirements will have to be met by incremental and radical innovations related to powertrain, fuel and material.

"Traffic, in particular automobile traffic, threatens to become a victim of its own success", says a car manufacturer. In fact, the negative side of an improved mobility consists of severe consequences for the environment and the economy due to congestion problems. Technological innovations and efficient organisational systems will help to solve current and future problems.

In spite of remarkable achievements in vehicle and road safety over the last decades there still is a long way to go towards the future vision of an accident-free traffic. Active safety systems for collision avoidance represent therefore the most efficient safety systems for the future. Better driver information and assistance on the driving conditions are part of the concept.

In all three of the mentioned areas – environment, mobility and safety – microsystems are playing a key role. More than in the past, however, integration aspects are becoming of increasing importance:

Microsystems have to be seen as part of an integral concept of a communicating and acting system embedded in new architectures and in structures of car-to-car and/or car-to-road communication.

This book published in occasion of the International Forum of Advanced Microsystems for Automotive Applications 2004 reflects the state-of-the-art of automotive R&D particularly in the area of automotive safety. By that it follows the priorities set by the European automotive industry and public authorities as for example the European Commission or the German Federal Ministry for Education and Research.

In the safety context a major role will play this year vision systems, emergency breaking and tire pressure measurement. Given the importance of integration aspects the networked vehicle issue will not be missing.

My explicit thanks go to the authors for their valuable contributions to this publication and to the members of the Honorary and Steering Committee for their commitment and support. Particular thanks are addressed to Gloria Pellischeck, CLEPA, and Berthold Ulmer, Daimler Chrysler for the inspiring input we received.

I would like to thank the European Commission for their financial support through the Innovation Relay Centre Northern Germany as well as the governmental authorities in Berlin and Brandenburg. Thanks go also to the supporting organisations Investitionsbank Berlin, mstnews and Victorian Trade Commission, Australia.

Last but not least, I like to express my sincere thanks to the Innovation Relay Centre team at VDI/VDE-IT, particularly to Jasmin Mehrgan and Angelika Boskugel for preparing this book for publication, and Jürgen Valldorf, as the AMAA project manager.

Teltow/Berlin, March 2004

Wolfgang Gessner

Public Financers

Berlin Senate for Economics and Technology

European Commision

Ministry for Economics Brandenburg

Supporting Organisations

Investitionsbank Berlin (IBB)

State Government of Victoria, Australia

mstnews

Co-Organisators

European Council for Automotive R&D (EUCAR)

European Association of Automotive Suppliers (CLEPA)

Advanced driver assistance systems in Europe (ADASE)

Honorary Commitee

Domenico Bordone	President and CEO Magneti Marelli S.p.A., Italy
Günter Hertel	Vice President Research and Technology DaimlerChrysler AG, Germany
Rémi Kaiser	Director Technology and Quality Delphi Automotive Systems Europe, France
Gian C. Michellone	President and CEO Centro Ricerche FIAT, Italy
Karl-Thomas Neumann	Head of Electronic Strategy VW AG, Germany

Steering Commitee

Dr. Giancarlo Alessandretti	Centro Ricerche FIAT, Orbassano, Italy
Alexander Bodensohn	DaimlerChrysler AG, Frankfurt am Main, Germany
Wilhelm Bois	Audi AG, Ingolstadt, Germany
Serge Boverie	Siemens VDO Automotive, Toulouse, France
Roger Grace	Roger Grace Associates, San Francisco, USA
Henrik Jakobsen	SensoNor A.S., Horten, Norway
Horst Kornemann	Continental Automotive Systems, Frankfurt am Main, Germany
Hannu Laatikainen	VTI Technologies Oy, Vantaa, Finland
Dr. Peter Lidén	AB Volvo., Göteborg, Sweden
Dr. Torsten Mehlhorn	Investitionsbank Berlin, Berlin, Germany
Dr. Roland Müller-Fiedler	Robert Bosch GmbH, Stuttgart, Germany
Paul Mulvanny	QinetiQ Ltd., Farnborough, UK
Dr. Andy Noble	Ricardo Consulting Engineers Ltd., Shoreham-by-Sea, UK
Gloria Pellischek	Clepa, Brussels, Belgium
David B. Rich	Delphi Delco Electronics Systems, Kokomo, USA
Dr. Detlef E. Ricken	Delphi Delco Electronics Europe GmbH, Rüsselsheim, Germany
Jean-Paul Rouet	Johnson Controls, Pontoise, France
Christian Rousseau	Renault S.A., Guyancourt, France
Ernst Schmidt	BMW AG, Munich, Germany
John P. Schuster	Motorola Inc., Northbrook Illinois, USA
Bob Sulouff	Analog Devices Inc., Cambridge, USA
Dr. Gerd Teepe	Motorola GmbH, Munich, Germany
Berthold Ulmer	DaimlerChrysler AG, Brussels, Belgium
Dr. Jürgen Valldorf	VDI/VDE-IT, Teltow, Germany
Egon Vetter	Ceramet Technologies, Melbourne, Australia
Dr. Matthias Werner	Deutsche Bank AG, Berlin, Germany
Arnold van Zyl	EUCAR, Brussels, Belgium

Conference chair:

Wolfgang Gessner	VDI/VDE-IT, Teltow, Germany

Table of Contents

Introduction

Safety

Powertrain

Comfort and HMI

Networked Vehicle

Appendices

Introduction

Bridging the Gap Between Technology Development and Market Implementation for Vehicle Safety Systems

F. Minarini, European Commission

Abstract

With its 375 million road users, mobility and transport is a concern for citizens throughout Europe. Modern society depends on mobility as an efficient transport system is the engine of our economy. The transport sector employs more than 10 million people with expenditure of more than 10% of the Gross Domestic Product (GDP) in Europe. The automotive sector manufactures about 17 million vehicles per year and employs with its suppliers close to 2 million people in Europe, with a turnover of 452 billion € world-wide [1]. The automotive telematic market, comprising of sales of platforms and services is experiencing rapid growth in market penetration, and according to some market studies, it will reach an annual revenue as high as 8,5 billion € in Europe in 2007, up from 1 billion € in 2000.

In the year 2001 there have been 1.300.000 accidents in Europe that caused 40.000 fatalities and 1.700.000 injuries, at an estimated cost of 160 billion €, or 2% of the GDP. On a personal level, these accident figures translate to one third of us being injured in an accident at some point of our lives. The psychological damage to the victims and their families cannot even be estimated.

Thanks largely to the efforts of the industry, vehicles today are inherently safer, cleaner and more recyclable than before. Thanks to improvements in the crash-worthiness of the vehicles, safety belts, ABS and other inventions, the vehicles are now four times safer than in 1970; this has largely contributed to the reducing by 50% the number of deaths in EU15 from 1970, while the traffic volumes have tripled during the same period. But the societal costs of road transport are still far too high, and new goals needed to be set. This is why, in its White Paper "European Transport Policy for 2010: Time to Decide", for the first time the European Commission has set an ambitious target to reduce by half the number of road fatalities by 2010 [2].

1 From Passive Safety to Active Safety and Accident Prevention

During the last decade, the European Union, Member States and the automotive industry have been actively involved in improving road safety through both accident prevention and injury reduction. Most of the accident prevention measures have focused on the driver, while the measures to reduce the consequences of an accident have primarily focused on the vehicle, through improved passive safety such as crashworthiness, seatbelts airbags and conventional active safety systems such as braking and lighting. Passive Safety measures have been proven to be a very effective method for the reduction of the car accident trauma.

These combined actions have contributed to the continuous reduction of the number of fatalities on European roads. The number of deaths in EU15 has halved from 1970, while the traffic volumes have tripled during the same period [3]. Nevertheless, the number of road accidents and the number of road victims are still unacceptably high in the European Union. Furthermore, the contribution of many of these "conventional" safety measures is reaching its limits, and further improvements in safety by these measures are becoming more and more difficult to achieve at a reasonable cost. This is why the in-vehicle passive safety has to be complemented by introducing on the markets more advanced in-vehicle and co-operative active safety systems.

2 Towards a New Generation of Intelligent Safety Systems Through Research

During the last decade in Europe both the industry and the public sector have invested heavily in Research and Development (RTD) in the use of Information and Communications Technologies (ICT) in in-vehicle safety technologies and subsystems. While the bulk of the work has been done by the industry, the EU's Research Programmes have contributed in realising leading edge technologies, systems and applications.

The EU has played a role in this research since the DRIVE Programme in 1988. Under the EU's Fourth Framework Programme for Research, Technological Development and Demonstration (1994-1998), the Telematic Applications Programme, helped in realising leading-edge systems and applications. The Information Society Technologies (IST) programme [4] continues research in technologies and applications systems aiming at safer, cleaner and more efficient transport, with specific focus on intelligent safety and Advanced Driver Assistance Systems and supporting technologies. The Intelligent Vehicle cluster of the IST Programme has over 40 projects, with total budget of over 150 million € and Community contribution of over 80 million €. Research into safer vehicles and infrastructures as well as accident databases and human behaviour in road transport is also undertaken in the Competitive and Sustainable Growth programme (5th Framework Programme) [5].

The development of appropriate sensors, actuators and processors, has already permitted wide spread implementation of systems, which help the driver to maintain control of the vehicle even when he has exceeded its "normal" limits of handling. Examples of such systems are Anti-lock Braking Systems (ABS) and Electronic Stability Programme (ESP). The latter one for example detects the onset of a slide and automatically applies the brakes to individual wheels to correct the slide and prevent spinning. ESP is now optional or even standard on many current passenger cars. Statistics from one car manufacturer shows that in 2001 there was a 4% reduction of accidents, compared to the year before, that could be directly attributed to ESP. The reduction of roll-over accidents thanks to ESP have also been estimated to be 12%.

The further development of intelligent active safety systems will require substantial RTD efforts. The current Sixth Framework Programme for Research and Technological Development (2002-2006) is offering new funding opportunities for RTD in intelligent integrated safety systems including accidentology and Advanced Driver Assistance Systems and technologies. The new instrument, Integrated Project, which is designed to generate the knowledge required to address major societal challenges, holds the potential to especially suitable for research in this area. Integrated Projects are intended to build a critical mass of activities and resources needed for achieving ambitious, clearly defined scientific and technological objectives. In the area of integrated safety systems, for example, an Integrated Project called PREVENT will start in February 2004. This project regroups more than 50 partners for a total budget of almost 80 million € with the European Commission contributing with 30 million € for a duration of 4 years.

3 Intelligent Vehicle Safety Systems

We know that almost 95% of the accidents are at least partly due to the human factor. In almost three-quarters of the cases the human behaviour is solely to blame. This apparent mismatch between driver skills and situation complexity can be addressed by improvements in three factors: the driver (education and training); the environment (intelligent infrastructure) and the vehicle (in-vehicle safety systems).

Intelligent Vehicle Safety Systems use Information and Communications Technologies for providing solutions for improving road safety in particular in the pre-crash phase when the accident can still be avoided or at least its severity significantly reduced. With these systems, which can operate either autonomously on-board the vehicle, or be based on vehicle-to-vehicle or vehicle-to-infrastructure communication (co-operative systems), the number of accidents and their severity can be reduced, leading equally to a reduction of the number of fatalities and injuries.

Collisions during lane changes and involuntary lane departure, for example are two of the most important causes of accidents. This problem requires in-vehicle technology to help detect and warn drivers of vehicles in adjacent lanes or when the vehicle is about to unintentionally depart from the lane. According to the National Highway Traffic Safety Agency (NHTSA) in the US, lane change and merge collisions could be cut in half by new technologies. In Europe, a Dutch study expects a reduction of 37% in all side impact collisions and a reduction of 24% of the single vehicle accidents due to lane departure crash avoidance.

Location-enhanced emergency calls like in-vehicle e-Call have their primary benefit to society of saving lives and in offering an increased sense of security. This is achieved by improved call routing obtaining faster and improved information for dispatching relevant resources, and most importantly improved information to locate the caller. What is of paramount importance here is that the relevant resources are delivered to the person in need as soon as possible, this can save up to 10% of the fatalities.

4   eSafety

The potential contribution of the introduction of Intelligent Vehicle Safety Systems for enhancing road safety and security has already been demonstrated by the industry in a number of European research and technological development (RTD) projects. However, to realise the potential benefits, the new systems have to be widely deployed in the marketplace and the existing gap between systems development and the market implementation of these systems has to be drastically reduced. Accelerating the development and deployment Intelligent Vehicle Safety Systems is the main objective of the eSafety initiative that was established during the High Level Meeting on 25 April 2002. The eSafety initiative is a public/private partnership that it has been created together by the European Commission and the industry mainly represented through ACEA and ERTICO. The actual work started in the High-level group when the e-Safety Working Group was officially established.

In November 2002, the Working Group published its Final Report, which the 2nd eSafety High-Level meeting endorsed as the basis for further actions.

This Final Report contains 28 recommendations for the European Commission, the Member States, road and safety authorities, automotive industry, service providers, user clubs, insurance industry and other stakeholders. The recommendations aim at improving road safety by integrated safety systems that use advanced ICT for providing new, intelligent solutions which address together the involvement of and interaction between the driver, the vehicle and the road environment.

Acting upon these recommendations, the Commission organised the first plenary meeting of the eSafety Forum in Brussels on April 22, 2003. The Forum brought together 150 representatives of the European automotive and telecommunication industries, providers of intelligent transport systems, infrastructure operators, public authorities and the Commission. Following the lines set out by its four preparatory working groups, the forum adopted recommendations on how to implement eCall (in-vehicle emergency calls), Accident Causation Data, a better Human-Machine Interaction and a stimulating Business rationale, and set up four new Working Groups covering International Cooperation, Real-time Traffic and Travel Information, Research and Development priorities and Road Maps. The second eSafety Forum took place in Madrid on November 17, 2003 on the occasion of the ITS (Intelligent Transport Systems and Services) World Congress.

The Commission has also adopted a Communication on "Information and Communications Technologies for Safe and Intelligent Vehicles". This Communication lays down a set of measures for supporting the industry in developing safer and more intelligent vehicles and enabling their rapid market introduction. The proposed actions fall into three categories: (I) Promoting Intelligent Vehicle Safety Systems, (II) Adapting the Regulatory and Standardisation Provisions and (III) Removing the Societal and Business Obstacles.

As the car park of vehicles with telematic grows, the market will shift towards services, further integrating the automotive market with two other key industrial sectors in Europe: Mobile Communications and Information Technology. As a whole, Information and Communications Technologies (ICT) holds the potential to play a key role in the convergence of these sectors towards the Intelligent Vehicles of the future that through the eSafety initiative and the commitment of the Industry and the public bodies like for example the European Commission and the Member States, can contribute to the set target of reducing road fatalities by 50% by 2010.

References

[1]   Sources: ACEA and Eurostat, 2001
[2]   White Paper on European Transport Policy for 2010, adopted by the Commission in September 2001
[3]   http://europa.eu.int/comm/transport/home/care/index_en.htm
[4]   The Information Society Technologies Programme is part of the European Union's Fifth Framework Programme for research and technological development (RTD), covering the period 1998-2002
[5]   See www.europa.eu.int/comm/research/growth/gcc/menu-researchthemes.html

Fabrizio Minarini
European Commission
Information Society Directorate General
Avenue de Beaulieu 31
1049 Brussels
Belgium
fabrizio.minarini@cec.eu.int

Safety

Monolithic Accelerometer for 3D Measurements

T. Lehtonen, J. Thurau, VTI Technologies Oy

Abstract

Bulk micromachining low-g accelerometers for single axis measurements are today's state of the art technology for measuring low-g
range signals in automotive, medical and other applications – also
referred as "human scale motion". The latest trend in silicon micro-
machining is smoothening the border between bulk and surface
micromachining. A combination of SOI (Silicon On Insulator) wafers
and DRIE (Deep Reactive Ion Etching) as new technologies is leading to smaller and more cost efficient micromachined inertial sensors. A novel capacitive 3-axis accelerometer element is the first
example of this combination by VTI Technologies. The key task was
to overcome the physical limitation to measure a 3-dimensional
acceleration signal with a 2-dimensional silicon element that suits to
modern packaging constraints. It will be shown how the new sensor
element is working and that the similarity of the manufacturing
process with today's mass produced VTI single axis accelerometer,
will lead to a fast commercialization of the novel 3-axis design.
Additionally the utilization of this technology platform for further
projects will be sketched.

1 Starting Position

Automotive inertial sensor road maps show a six degrees-of-freedom inertial
measuring unit, where a three-axis accelerometer is needed for measuring linear motion. The three-axis accelerometer is also expected to open up new markets for accelerometer based motion sensing, navigation and inclination measurement in portable electronic devices [1]. Several previous attempts to design
and manufacture a three-axis accelerometer are reported in the literature
[2-5]. VTI Technologies has been developing and producing single axis low-g
bulk micromachined capacitive accelerometers since the early 90's and has
achieved a leading position with over 10 million annual production [6, 7]. Great
deal of recent research and development work has been done to develop a new
concept for a single element multiple axis accelerometer.

The approach described in this paper is a 3-axis accelerometer which requires a high sensitivity system consisting of springs and a proof mass or several masses, whose position is observed capacitively and further mapped to acceleration information by an electrical circuit. Several requirements for the sensor element are existing such as: mass movement, conversion to an electrical signal and the mapping algorithm should yield a high signal to noise ratio and a linear acceleration response with negligible cross axis sensitivity. In the ideal case each measured axis should be mechanically equal in sensitivity and frequency response – or even better, freely adjustable over a wide range. The next important specifications are temperature dependence and shock resistance. It isn't less important that the accelerometer should be manufacturable based on feasible low cost processes, have small size and should enable low cost interconnecting and packaging.

VTI inertial sensor technology is based on bulk micromachining using six inch wafers to meet all above mentioned requirements. Bulk micromachining technology enables the formation of large inertial masses which have high sensitivity, reliability and resistance to stiction effects. Five layer sensor structure enables to achieve large capacitances using sensor element area effectively. 35% of the area is covered by the capacitor electrode. An effective wafer area usage results in small size and low cost.

2 Operation Principle

The structure is based on multiple proof masses suspended asymmetrically with torsion springs attached close to the surface of the silicon wafer. An inertial mass with relatively high thickness attached in this way is sensitive in vertical and horizontal directions of the wafer. The resultant sensing direction depends on the location of the proof mass center of gravity with respect to the suspension points. The horizontal surfaces of the mass and metal electrodes on the capping wafers on both sides of the mass form a differential capacitor system. When the mass is excited by acceleration force, it will rotate. The movement changes the electrode distances and thus the capacitances. Figure 1 illustrates a single mass operation and fixed electrodes. Solid line in figure 1b indicates the mass position in the rest and dashed line mass position when excited in sensing direction.

A minimum of three spring-proof mass systems is needed for a 3-axis accelerometer. By measuring the positions of the proof masses capacitively, the three orthogonal components of the acceleration vector can be calculated as linear combinations of the position signals from several proof mass-electrode systems.

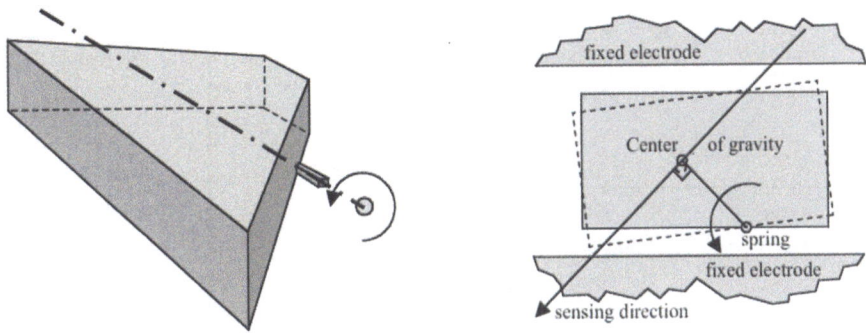

Fig. 1.   Single mass function principle (left), as cross section (right).

VTI's 3-axis capacitive accelerometer consists of four triangular spring-proof mass systems located at 90° angles from each other. Masses form totally eight capacitors and four differential capacitor pairs with fixed electrodes. Figure 2 shows the spring-proof mass systems in a Cartesian co-ordinate system as well as a photo of the spring system made from one of the prototypes.

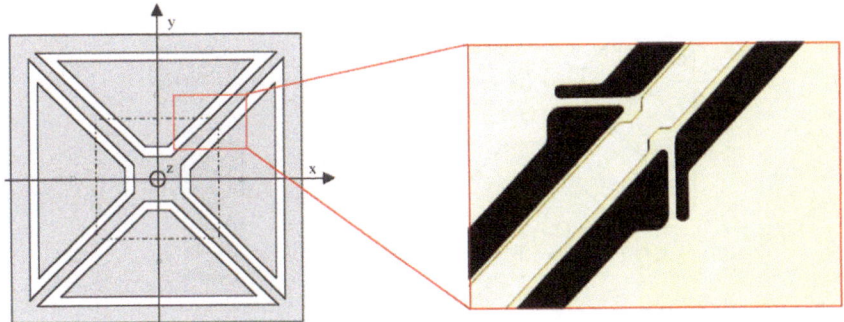

Fig. 2.   Complete spring-proof mass systems with spring detail.

When acceleration load is applied on sensor, the masses will rotate accordingly. Figure 3a shows an example of mass response when acceleration load is purely in positive x-direction. Acceleration makes masses 1 and 2 rotating clockwise resulting in decreasing electrode gap for electrode 2 and 3. Correspondingly electrode gap for electrodes 1 and 4 will increase. Masses 3 and 4 are not affected and capacitance change is not recognised. Figure 3b shows the situation when acceleration in positive z-axis direction is applied. As the result, masses 1 and 3 will rotate counter clockwise and masses 2 and 4 clockwise. The acceleration resultant can be calculated from mass signals.

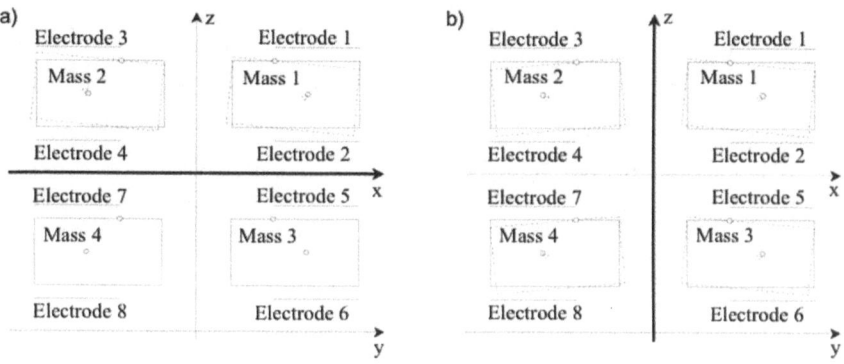

Fig. 3. Mass modes when excited in a) x-direction and b) z-direction.

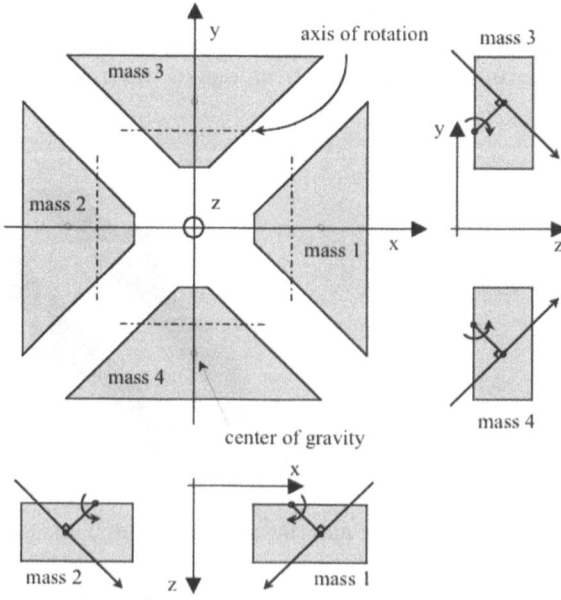

Fig. 4. Projected resulting force vectors.

The concept of four inertial masses enables very similar linear responses for all three measurement axes. Measurement for X- and Y-axis can be adjusted independently for different mechanical sensitivities and frequency bands. Z-axis measurement properties are dependent on X- and Y-axes properties, but still have equally high performance.

Figure 4 shows the same system again but now with projection in two planes. A set of each two of four resulting force direction vectors is perpendicular to each other orientated 45° to the plane. The 45° angle out of the plane was chosen as the best result after various optimisations within the given degree of freedom.

3   Sensor Element

For best system integration it is necessary to realise a very compact sensor element. The inertial mass-spring systems are suspended to the frame of the sensor, which encloses each mass to its own cavity according to figure 2. These four cavities add up to a square of 2.65 mm by 2.65 mm. The structural wafer with proof masses is enclosed between two capping wafers consisting of silicon blocks embedded in a glass matrix. The electrode metallization on the surfaces of the capping wafers functions as a fixed electrode. Electrodes are further connected to bonding pads via silicon blocks and electrical feed throughs. In the fabricated prototypes the total thickness of all three layers is 1.5 mm. The accelerometer prototype is shown in figure 5.

The sensor layout was dedicated for low-g measurement range incorporating high accuracy and stability. The prototype sensor has a large active capacitance value of 3.5 pF and capacitance dynamics of 10% per g for ±2 g version. The sensor frequency response is dependent on cavity interior pressure, capacitor gap height and mass surface shape. For the prototype sensor the -3 dB frequency has been adjusted to 200 Hz. Estimated mechanical noise due to the damping is 5 µg/sqrt(Hz).

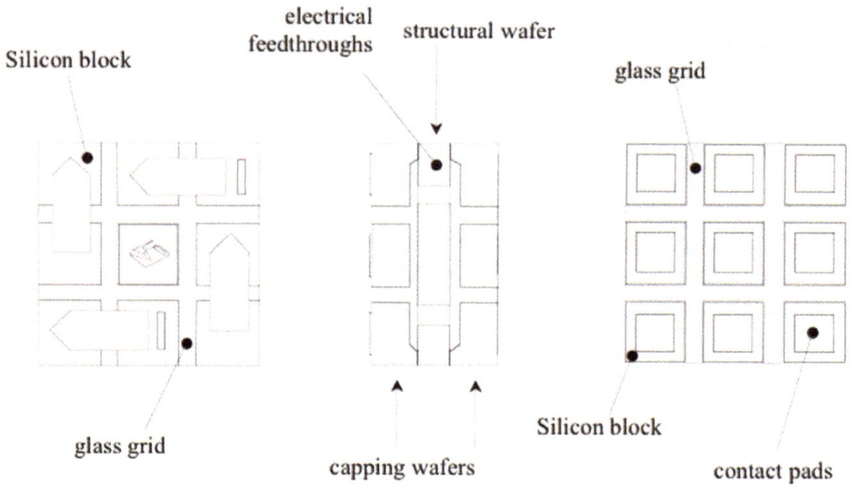

Silicon block

electrical
feedthroughs

structural wafer

glass grid

glass grid

capping wafers

Silicon block

contact pads

Fig. 5.   3-axis accelerometer prototype.

The geometrical structure is highly symmetrical and has a rigid frame (figure 2), which extends over the whole sensor area. This rugged structure protects the capacitors from major deformations by interconnection, temperature loads or other external impact due to mounting. Minor temperature dependency will exist due to thermal expansion mismatch between silicon and glass and due to the dielectric coefficient of the glass. Temperature dependency is cancelled to a large degree by the differential capacitance arrangement and it is similar in all four differential capacitors due to high level of structural symmetry. The sensor's shock resistance is improved by several design features. For example protective bumpers limit mass deflection and spring ends are rounded preventing high stress concentrations to occur.

Fig. 6.   Model and prototype element hardware on 1cent coin

The sensor element has many possible interconnection and packaging alternatives. It can be die attached to a substrate or in a housing and wire bonded. There is also possibility to directly re-flow solder the accelerometer as an SMD component. Flip-chip bonding with proper interconnection bumps would be another option that is under discussion.

4   Manufacturing Process

A new degree of effective use of silicon area was achieved by implementing Deep Reactive Ion Etching (DRIE). This process enables completely vertical cuts through the silicon wafer without limitations of silicon crystal orientation. Space used for mass separation can be minimised and shapes can be freely designed in 2D taking only into account the optimum sensor performance. In addition to the DRIE, SOI (Silicon On Insulator) wafers have been introduced for this concept. The SOI insulator layer is used here to achieve an efficient etch-stop with double-sided DRIE. The thickness of mechanical torsion springs is defined accurately by the initial device layer (isolator thickness). The sensor element layout was designed to be robust so that even with non-ideal processed DRIE trenches the sensor performance will still meet sensor specification. This is due to the fact that the electrical important surfaces (capacitor plates) are perpendicular to the trench edge direction.

The 3-axis accelerometer is partly based on technology choices similar to VTI's single axis accelerometers: A 5-layer structure consisting of single crystal silicon and borosilicate glass is prepared separately on one structural silicon wafer and two capping wafers. Vertical glass isolations used in capping wafers combine thick glass isolation for multiple feed throughs to the hermetic sensor interior.

The fixed electrodes are formed by thin film process on both capping wafers. The electrodes are connected to the outer contacts by glass feedthrough structures. The thick glass layer on both capping wafers will be bonded anodically to structural inner wafer with its spring mass system to a hermetic sealed sensor element. The purely mechanical frequency response of the sensor element is adjusted just by the amount of remaining gas fill in the element. Basically the final behaviour of the cell is similar to today's VTI sensors in series production. The parasitic capacitances are optimised to extremely low values for best sensor performance.

5 Signal Conditioning

The stable properties of the sensor element as the heart of the sensor system are the basis for excellent sensor system behaviour. To achieve the required performance the sensor interface has to be carefully designed also. The sensor signal measurement and processing requires many steps. Figure 6 shows a schematic signal flow from real acceleration to digital acceleration signal. Each differential capacitance pair is measured by an analog circuit and transformed to voltage. Four voltage signals are passed to a block for linear mapping to achieve three independent acceleration measurement output signals.

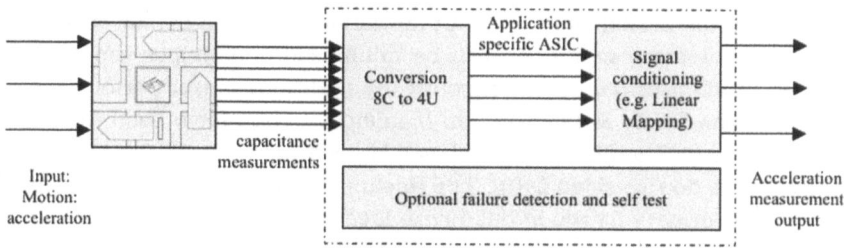

Fig. 7.   Schematic signal flow for 3-axis acceleration measurement.

For low g-sensors it is sufficient to perform calibration in earth gravity field due to the uniform behaviour of the sensor elements. The worst case approximation for summed axes mixing and linearity error is 0.1% full scale for ±2 g and 0.3% full scale for ±6 g. All the linearity and axis mixing calculations are based on accelerometer mechanical behaviour and assume ideal signal processing as well as ideal mounting of the sensor element.

The strength of the concept is that the sensor element and the ASIC are processed separately. This leads to the advantage that the signal conditioning could be adapted to application demand, like low power consumption, higher precision or various analog or digital output formats.

A failure detection functionality could be added since the system is overspecified with four masses so that by using that redundancy an unplausible status of the sensor could be recognised and flagged-up. Several different strategies are under discussion. The main target is to keep the ASIC complexity and size low so that cost targets are met. Therefore a straight forward strategy will be preferred.

6   First Measurement Results with 3-Axis Sensor Element

Prototypes of the sensors have been manufactured. Each of the eight capacitors has close to calculated active capacitance level, sensitivity and capacitance change with acceleration. Figure 8 illustrates the change of each capacitance when the accelerometer is rotated in earth's gravity field around positive y-axis in figure 4 starting from the -1 g position for z-axis. Each capacitance reading is a sum of active and parallel parasite capacitance.

It needs to be taken into consideration that those output signals were direct readings from the sensor element and not subject to signal conditioning. The signal conditioning needs to take care of part to part variation in terms of offset and sensitivity adaptations for each single mass.

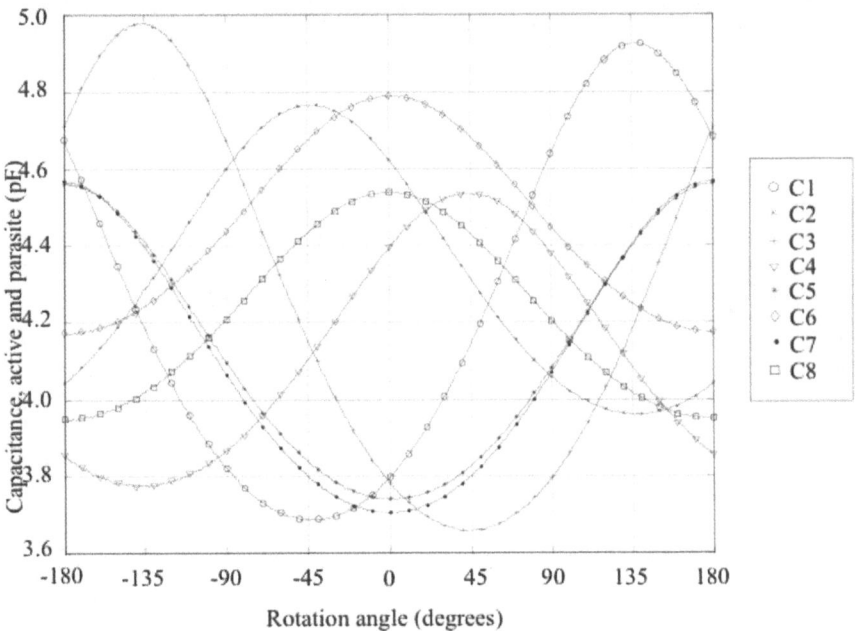

Fig. 8.   8 capacity reading before signal conditioning.

7 Packaging

Different types of applications are underlying different requirements not only for the signal conditioning but also for the packaging. The 3-axis element will be combined with different type of housings for consumer, industrial and automotive application. The concept that is mainly addressed to the automotive market might look similar as shown in figure 9. The idea is to have a common PCB layout or footprint for VTI's new single axis SCA800 sensor as well as for the 3-axis version SCA3100 in a slightly bigger housing but with a similar pinout. This would lead to optional use of an appropriate single-axis or multi-axis product. A similar solution is already realised with today's SCA610/620 single axis family in relation to SCA1000 family hence this family needs approximately 50% more space on the PCB as the next generation of products.

These housings will fulfil harsh automotive environmental requirements, like temperature range from −40°C to 125°C with an appropriate robustness against thermal shock. Both housings are still in draft status so that a final outline will be presented later in the projects.

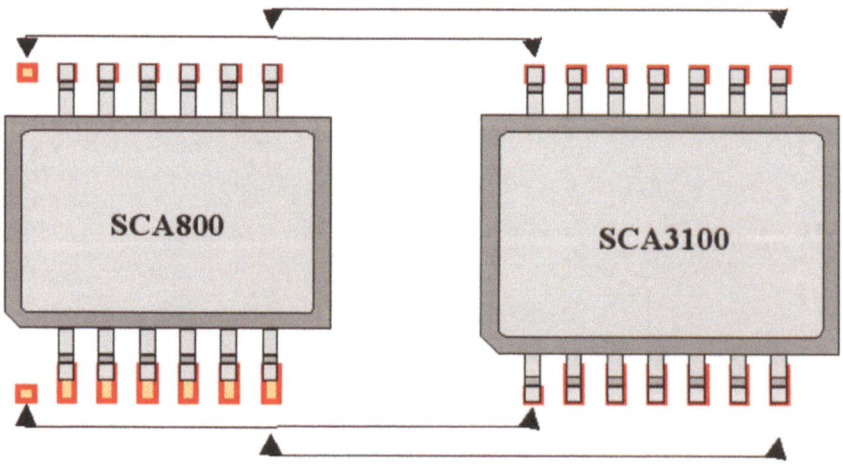

Fig. 9.   Upgrade compatible 1axis SCA800 and 3axis SCA3100 (draft).

8 Outlook

It was demonstrated that the new 3-axis accelerometer concept fulfils the requirements set for the sensor element: It has a very high sensitivity and it is robust. It shows similar behaviour and performance regarding linearity and over temperature stability as it is known from VTI's accelerometer in series production as of today. The manufacturing process is feasible and requires additionally only one major new process – the Deep Reactive Ion Edging – to the existing high volume processes. The sensor element has a small size and it is inherently fully hermetic. The rugged design enables to use wire bonding connection as well as flip-chipping or even SMT for interconnection. It is seen as a major advantage in terms of possible system performance that the novel capacitive accelerometer has large capacitance levels and sensitivity compared with other capacitive accelerometer elements [3, 4].

Beneath the development of application specific housings as well as signal conditioning circuitry a further optimisation of the sensor element towards lower height with thinner structures is currently ongoing. It is targeted for applications were the height of the sensor element is seen as critical.

References

[1] MEMS-IMU Based Pedestrian Navigator for Handheld Devices: J. Käppi, University of Technology, Finland; J. Syrjärinne, Nokia Mobile Phones, Finland; J. Saarinen, Tampere University of Technology Finland, The Institute of Navigation GPS 2001, Salt Lake City, USA

[2] Sebastian Butefisch, Axel Scoft, Stephanus Buttgenbach,'three-Axes Monolithic Silicon low-g Accelerometer', Journal of Microelectromechanical Systems, Vol. 9, No. 4, December 2000

[3] R. Puers, S. Reyntjens,'Design and processing experiments of a new miniturized capacitive triaxial accelerometer', Sensors and Actuators A 68, 1998

[4] Gang Li, Zhihong Li, Congshun Wang, Yilong Hao, Ting Li, Dacheng Zhang and Guoying Wu, 'Design and fabrication of a highly symmetrical capacitive triaxial accelerometer', Journal of Micromechanics and Microengineering, 11, 2001

[5] M. A. Lemkin, M. A. Ortiz, N. Wongkomet, B. E. Boser, and J. H. Smith, "A 3-axis surface micromachined SD accelerometer," in ISSCC Dig. Tech. Papers, pp. 202–203, Feb. 1997

[6] Heikki Kuisma, 'Inertial Sensors for Automotive Applications', Transducers '01, Munich, Germany 2001

[7] Heikki Kuisma, Tapani Ryhänen, Juha Lahdenperä, Eero Punkka, Sami Ruotsalainen, Teuvo Sillanpää, Heikki Seppä,'A Bulk macromachined Silicon Angular Rate Sensor', Transducers '97, Chicago 1997

[8] T. Lehtonen, "A Novel 3-axis Capacitive Accelerometer Concept", Sensor Nuremberg, Mai 2003

Tuomo Lehtonen, Jens Thurau
VTI Technologies Oy
P.O. Box 27
01621 Vantaa
Finland
Tuomo.Lehtonen@VTI.FI; Jens.Thurau@VTI.FI

Keywords: 3axis accelerometer, bulk micromachining, DRIE etching, signal conditioning

Highly Integrated Tire Pressure Monitoring Solutions

M. Osajda, Motorola

Abstract

This paper is presenting a new tire pressure monitoring (TPM) sensor that offers a high level of functional integration by combining a 0.8 µm CMOS surface micro machined capacitive pressure sensing element with a 0.35 µm SmartMOS™ ASIC into a single small outline 16 pin package. The pressure sensor feature low voltage, low power operation, and include an on-chip CMOS radio frequency (RF) transmitter, a low frequency (LF) receiver, a temperature sensor, a voltage sensor as well as embedded motion sensing capability. The data processing is performed by the high logic gate density programmable state machine and associated non-volatile memory containing the device settings. Several mission profiles and operating modes can be programmed to comply either with the US Federal Motor Vehicle Safety Standard 138 or with any other existing car manufacturer requirements throughout the world.

1 Summary of US Legal Requirements

In May 2002, the US National Highway Traffic Safety Administration (NHTSA) officially released the Federal Motor Vehicle Safety Standard 138 (FMVSS138), Tire Pressure Monitoring Systems, Controls and Displays that requires the installation of tire pressure monitoring systems to warn the driver when a tire is significantly under inflated.

Application - Passenger cars, trucks, multipurpose passenger vehicles, and buses with a gross vehicle weight rating of 10000 pounds (~4536 kg) or less, except those vehicles with dual wheels on an axle.

Compliance Option - Both Direct and Indirect system will be allowed during the phase-in period from November 2003 through October 2006 and with the following technical requirements. (Note: A US federal court decision dated August 6, 2003 will prevent the car manufacturer to implement indirect system as currently specified by NHTSA.

The agency has decided to revise the standard. At the date of writing this paper, the new standard was not released).

Indirect Measurement – Indirect TPM must warn the driver when the pressure in any single tire has fallen to 30% or more below the vehicle manufacturer's recommended cold inflation pressure for the tires, or a minimum level of pressure specified in the standard, whichever pressure is higher.

Direct Measurement - Direct TPM must warn the driver when the pressure in any single tire or in each tire in any combination of tires, up to a total of four tires, has fallen to 25% or more below the vehicle manufacturer's recommended cold inflation pressure for the tires, or a minimum level of pressure specified in the standard, whichever pressure is higher.

Phase in Schedule – Specified in percentages of vehicles sold in the USA (table 1).

November 2003 through October 2004	min 10%
November 2004 through October 2005	min. 35%
November 2005 through October 2006	min 65%
After November 2006	100%

Tab. 1. FMVSS138 USA Phase In schedule.

There are currently no legal requirements outside the USA, but thanks to the added safety provided by such a system, most of the car manufacturer around the world have decided to implement TPM on most of there new platforms.

The next paragraphs of this paper will present a solution dedicated to direct pressure monitoring.

2 A New Highly Integrated Tire Pressure Monitoring Sensor

2.1 Technical Requirements and Challenges

The implementation of the FMVSS138 in the USA has driven the automotive and semiconductor industry to a new level of integration and cost reduction rarely achieved in this industry. In order to respond to this new requirement, Motorola has developed a new device with all the required TPM functionalities embedded in a single surface mount package.

2.2 TPM Required Set of Features

The following feature set is what is commonly required by the TPM system manufacturer: Absolute pressure measurement, temperature measurement, voltage measurement, motion detection, embedded signal conditioning and signal processing, access to multiple datagram format, possibility to communicate with the device via an LF channel, embedded RF transmitter, ultra low power consumption and cost effective.

The challenge for Motorola was to create such a device by combining the appropriate technologies in a single package, and being able to produce it in very large volumes. The following approach has been taken.

2.3 The SmartMOS™ ASIC, the Brain of the System

The basic functions to be performed in a TPM module are relatively straight forward: measuring pressure, measuring temperature, and transmitting both data and a serial number. But to support this basic function there are several subsystems or modules that have their own distinct events to analyze. TPM system enhancements may also bring other modules into consideration such as measuring the battery voltage or ability to communicate bi-directionally with the TPM system. And finally, some modules may be introduced with the sole intent to reduce power consumption. Figure 1 shows a typical architecture of a wheel module with the required functions.

Pressure Measure (1) – The event must power up the sensing element, convert its variation to a voltage, store that result using a sample and hold capacitor and then shut down.

Temperature Measure (2) - The event must power up the sensing element, convert its variation to a voltage, store that result using a sample and hold capacitor and then shut down.

Voltage Measure (3) - The voltage measurement is primarily to monitor the battery. It is done using a voltage threshold comparator for the minimum battery voltage. There must also be a fairly accurate voltage reference. This event must power up the voltage reference and the comparator, store the result and then shut down. There will also be a requirement to perform this voltage measurement during both the highest current event (usually the RF transmission) and during any event requiring the highest battery voltage (usually an analog measurement).

A/D Conversion (4) – If any of the prior measurements create an analog voltage output, then an A/D converter will be required to convert the voltage to a digital code for further processing. This event will be to power up the A/D converter, convert a stored voltage, store the resulting data, and then shut down.

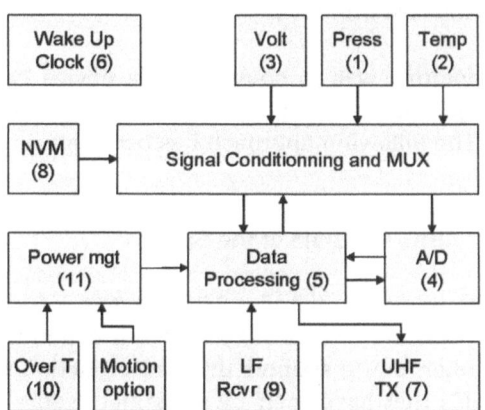

Fig. 1.   Wheel module architecture.

Data Processing (5) – The data collected from the above measurements must be analyzed according to a defined operational algorithm. The result will determine how the TPM will operate in the future and whether a transmission is to be made. The data processing will also format the data for the RF transmission. The power consumed by such digital logic will depend on a very small standby current term and a speed-related term that is linearly related to its clocking speed. Further, each data process can be separated out into individual events such as read data, calculate data, format data, send data, etc.

Wake Up (6) – The wheel sensor will normally be in a standby mode waiting for a need to measure and process data. This wake-up function is usually a very low power oscillator that outputs a pulse every 3 to 6 seconds. This event will be one of the long time standby current events that may exist for the complete life of the battery.

RF Transmitter (7) – The RF transmitter module is a combination of some digital processes and the actual RF power transmitter. This is the highest current consumption module and creates the highest current pulse. However, while transmitting the operational voltage may be less than required for the measurement modules. Therefore the battery cut-off voltage will be different from that for the other modules. In all cases, a minimal number of RF transmissions are the most important feature of the operational algorithm. A confounding problem with the RF transmitter is how to handle the possibility of simultaneous transmissions from multiple wheel sensors. This would result in data "collisions" causing the RF receiver to be unable to discern any data. The usual method to solve this issue is to randomize the transmission times and also send multiple randomized data words within any given transmission.

NVM Memory (8) – The non-volatile memory (NVM) module serves several purposes. One purpose of the NVM is to store trimming values to make each sensor meet the required accuracy for pressure and temperature measurements. One other purpose for the NVM is to store a unique serial code for each wheel sensor and to store the configuration bit of the device. The NVM may draw current for each programmed bit as well as some digital switching current during its access.

LF Receiver (9) – One method to identify and/or wake up the wheel sensor is to transmit a low frequency electro-magnetic signal to it. This is usually done at frequencies around 125 kHz using a transmitting coil mounted on the chassis or axle housing somewhere near each wheel and a small LF receiver coil within each wheel sensor. The LF receiver can also be used to trigger a given wheel sensor to transmit within a given time frame. In this way the chassis receiver can match up the wheel sensor's serial ID with the location that was triggered and provide tire localization information. A further use of the LF receiver would be to receive test and other commands from a closely placed inductive transmitting coil. This can provide diagnostics and fast identification of wheel sensors in either the factory or field.

Over-Temperature (10) – A concern with TPM is that extremely high temperatures may develop within the tire and wheel, which are beyond the normal operational range for the wheel sensor components. The major source of such high temperatures is the braking system that can transfer heat through the hub to the wheel rim. The concern is whether the wheel sensor may start to operate erratically above 125°C and either transmits erroneous data about the tire or starts transmitting continuously and therefore drains the battery unnecessarily. Some sort of over-temperature switch is needed to prevent data transmission above 125°C.

Power Management (11)– The most critical operational issue with TPM is keeping total power consumption to a minimum. Therefore, power management requires special attention. This could be accomplished by simply providing on/off enables in each module and controlling these enables with a small microcontroller unit (MCU). The drawback of this approach is that the MCU requires a number of instructions to manipulate the enables. A simple change of one control line or a logical AND or OR function may consume multiple central processing unit (CPU) bus cycles. If such power control sequences were controlled by logic the changes would be nearly instantaneous and consume much less power. Switching of a number of logic nodes (such as in an MCU bus) takes more power than switching a single NAND gate in a state machine.

Therefore as many tasks as possible should have their power sequenced using hardware logic in small state machines. The implementation of those small state machine is possible by using a process such as SmartMOS™, a process enabling the combination of analog, digital, NVM and RF functionality on a same die as shown in figure 2.

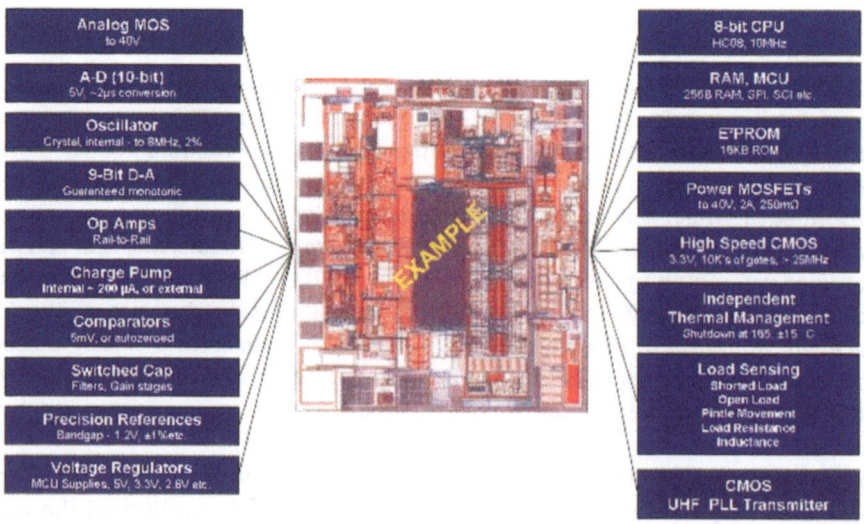

Fig. 2.   Capabilities of SmartMOS™.

2.4 The Surface Micromachined Sensing Element

The primary goals of the transducer design are the following:
1) Produce a sufficient base capacitance
2) Minimize pressure linearity errors
3) Maximize pressure sensitivity
4) Minimize the die area
5) Minimize temperature errors
6) Minimize the parasitic capacitance to the substrate.

Traditionally, there has always been a trade-off between items 2 and 3; greater sensitivity usually produces larger linearity errors. Although linearity errors can be improved with reduced pressure sensitivity, more circuit gain is required, and circuit noise and other errors can increase. Thus, there is a desire to have a high-pressure sensitivity and a low-pressure linearity error, and therefore, other approaches were investigated to reduce linearity errors.

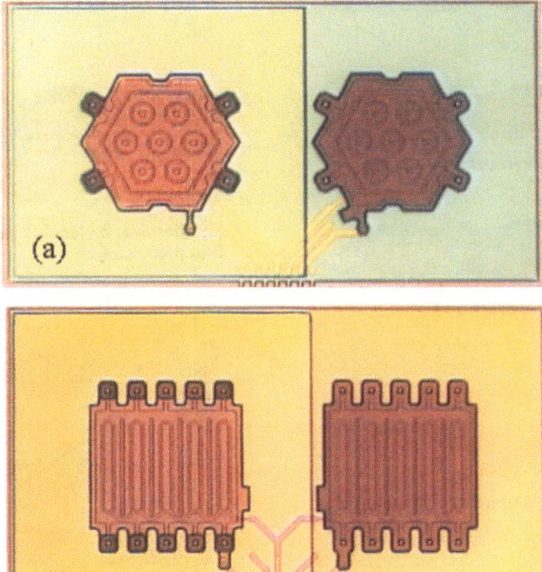

Fig. 3.   Transducer using an internal post diaphragm design(a).
Transducer using a rectangular diaphragm design (b).

Finite element simulations were used to evaluate candidate diaphragm shapes for uniformity of diaphragm deflection. In addition to previously used shapes, such as circles and squares, "internal post" configurations were evaluated. Internal post designs anchor the diaphragm both at its perimeter and at internal locations. Simulations showed that some configurations of internal posts could produce more uniform diaphragm deflection, and significantly reduce linearity error at a given sensitivity level. Figures 3(a) and 3(b) show photographs of an internal post design and a rectangular diaphragm design.

The methodology used in the insertion of the sensor module into the CMOS process was based on the requirements that there is no impact, either thermally or topographically, on the CMOS fabrication. This is done by insertion of the micromachining steps before the temperature sensitive process steps of the CMOS process. It was identified that the source/drain implants and the subsequent thermal processing were the temperature sensitive steps. In addition, the process integration was designed to reuse, as much as possible, the fabrication steps of the CMOS process. The cross-section of the CMOS and sensor devices is shown in figure 4 illustrating the key features of the process integration.

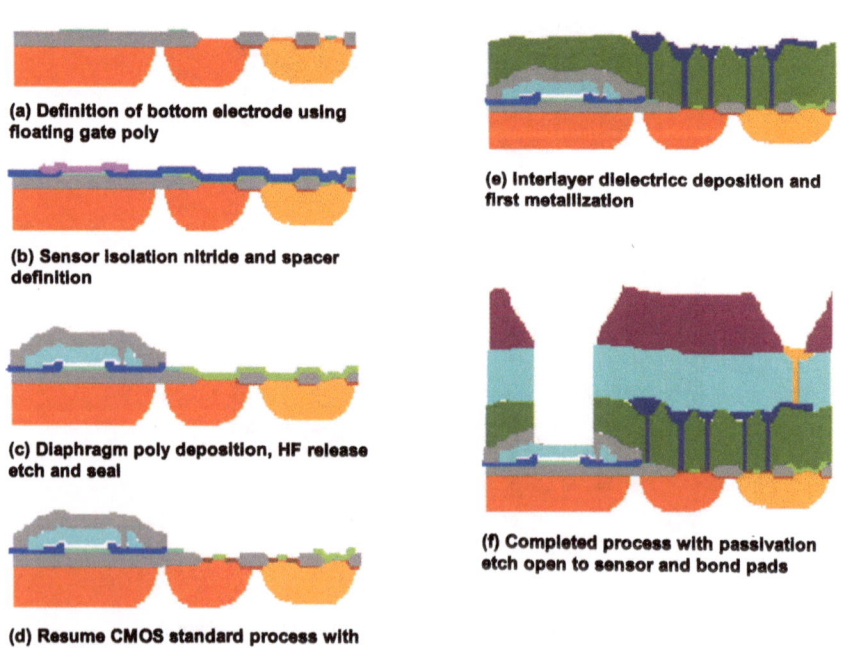

(a) Definition of bottom electrode using floating gate poly

(b) Sensor isolation nitride and spacer definition

(c) Diaphragm poly deposition, HF release etch and seal

(d) Resume CMOS standard process with Source/Drain formation

(e) Interlayer dielectricc deposition and first metallization

(f) Completed process with passivation etch open to sensor and bond pads

Fig. 4.   Selected cross-sections of CMOS integrated capacitive pressure sensor process flow.

The fabrication process starts with the standard CMOS fabrication process in which the active areas are defined using an isolation field oxide using a LOCOS process. In addition to defining the active area of the CMOS devices, the field oxide is used as the area where the sensor is to be formed. The next step is the formation of the floating gate of the EEPROM (note: EEPROM may be implemented either on the sensing element or on the ASIC, depending on the system partitioning). The required extension implant of the EEPROM is completed followed by the deposition of the thin tunnel oxide. The EEPROM floating gate is then deposited using a doped LPCVD polysilicon film. The floating gate polysilicon is patterned and etched so that it also forms the bottom fixed plate of the capacitive pressure sensor. The isolation of the fixed electrode to the substrate is provided by the CMOS field oxide isolation. The CMOS transistor gate oxide is then deposited followed by the gate polysilicon deposition.

Fig. 5.   CMOS Integrated pressure sensor.

At this stage, the CMOS fabrication is suspended and the sensor fabrication steps are inserted. The first step is the deposition of a layer of PECVD oxide to protect the CMOS gate Poly from future sensor processing. The next step is the removal of the PECVD oxide, gate polysilicon and the gate oxide from the sensor area. This is done in such a way that the bottom polysilicon plate is not affected by the etching process. After the gate oxide is removed, an isolation layer of low stress silicon rich nitride is deposited using an LPCVD process. The silicon rich nitride is chosen for its excellent etch resistance to HF based chemistries. The silicon nitride is used to isolate the top and bottom plates of the pressure sensor and also protect the CMOS area from subsequent micro-machining steps.

The sacrificial layer of the sensor is next deposited using a phosphorus-doped glass. The thickness of this layer defines the spacing between the bottom fixed electrode and the moveable diaphragm. The spacer layer is also patterned for transducer designs, which utilize internal support posts for improved pressure linearity.. The sensor diaphragm is then formed by deposition of a LPCVD poly-silicon layer. The polysilicon is then doped, patterned and etched to define the diaphragm.

The pressure sensitive diaphragm is then released by removing the sacrificial layer below the diaphragm polysilicon. Etch ports located at the periphery allow the HF based chemistry to completely remove the phosphorus-doped glass. The etch ports are then sealed using a line of sight sealing with a PECVD deposited oxide layer. The thickness of the PECVD oxide is chosen to ensure that the etch holes are plugged and the cavity between the fixed plate and the diaphragm is sealed at the deposition pressure of the PECVD oxide. This com-pletes the formation of the sensor module.

The next step is the resumption of the CMOS fabrication process. This is done by first removing the sealing oxide, isolation nitride and protective oxide over the CMOS device area. The CMOS device formation is then resumed with the formation of the source/drain regions, and the subsequent thermal treatment to provide the activation anneal and definition of the junctions. This thermal treatment also serves to activate the dopants and provide the stress relief anneal in the sensor polysilicon film.

The first interlayer dielectric layer is next deposited, patterned and etched to allow for the subsequent first metal layer to contact the various CMOS devices as well as provide the interconnect for the sensor to the circuit. The second interlayer dielectric layer is then deposited, patterned and etched to form vias for the second metal to contact the first metal with the routing required for the circuit blocks. Finally, the passivation layer is deposited over both the CMOS and sensor device area. The passivation layer is then patterned and etched to expose the bond pads. In addition, the passivation layer, second, and first dielectric layers are then removed over the pressure sensitive element but not from the reference sensor element. This exposes the polysilicon diaphragm of the sensitive element leaving it free to deflect under an applied pressure. The reference element has very little deflection since it has a very thick composite diaphragm. Both the pressure sensitive element and the reference elements are used by the circuit to generate the output of the sensor. This completes the description of the process integration of the capacitive pressure sensor with the CMOS devices. Figure 6 shows a die picture of a CMOS integrated pressure sensor.

3 The Combination of a Sensing Element and Dedicated ASIC in a Single Package

3.1 Block Diagram

Figure 6 shows the block diagram of the new sensor with all the previously described features integrated on two different chips: A surface micromachined sensing element and a SmartMOS™ ASIC, both packaged into a dedicated SOIC16 package. A third chip, more specifically a piezoelectric ceramic can be included in the package to provide an embedded motion detection (not shown and not further described in this paper).

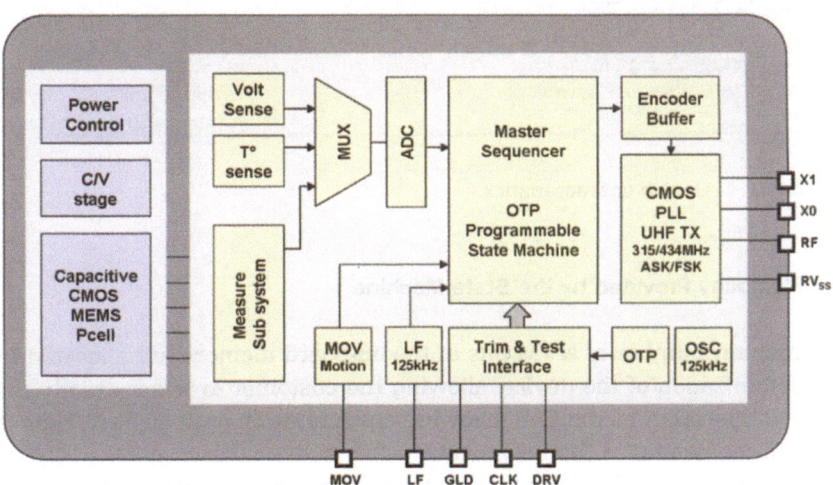

Fig. 6.   Block diagram of the new sensor.

3.2 Device Characteristics

Operation Range	
Supply Voltage	2.1 to 3.4 V
Continuous Temperature Range	-40 to 125°C
Temperature Excursion < 10 min	-40 to 150°C
Sensors Characteristics	
Pressure Range	50 to 450 kPa or 50 to 900 kPa abs.
Pressure Accurancy (*)	+/- 7 kPa to +/- 15 kPa
Temperature Range	-40 to +125°C
Temperature Accurancy	+/- 4°C
P, T and V resolution	8 bits
Communication Interface	
RF Data Encoding	Manchester or Bi-Phase
RF Modulation Schema	ASK or FSK
RF Baud Rate	4.8, 9.6 & 19..2 kbs
RF Output Power	-1 dBm or +5 dBm
LF Input Sensitivity	Up to 2 mVpp
LF Input frequency	125 kHz typ.
Power Consumption	
Standby (*)	1.1 uA typ
During Measurement (P,V or T) (*)	20 uA sec typ
During Transmission (9.6k) (*)	200 uA sec typ
Peak Current (*)	8 mA typ
Package	
Centrifugal Forcce Resistance	2000 g
Unpowered Shock Resistance	2000 g
Maximum Pressure (absolute)	1200 kPa
Package	Media Resistant, Integrated Air Filter

Tab. 2.   Device characteristics.

3.3 Flexibility Provided by the State Machine

A significant number of NVM bits in the on-board memory are allocated for the customization of the device, allowing the customer to select exactly the required operating mode. The following customization options have been currently implemented in the new sensor:

- ▶ 4 mission profiles, each defining a set of timing between the different states.
- ▶ 15 transmit mode in order to address the different receiver configurations.
- ▶ 2 modulation frequency bands to address the different RF regulation world wide.

- ▶ 2 RF output power options to optimize the power consumption per car platform.
- ▶ 2 datagram format options.
- ▶ 2 data frame bit length options (82 bit / 90 bit).
- ▶ 2 data encoding schemes, Manchester or Bi-phase.
- ▶ 2 data polarity options (normal / reversed).
- ▶ 2 LF sensitivities options, high and low.
- ▶ 2 averaging methods, to minimize noise.
- ▶ 4 delta pressure algorithms, defining the conditions to enter into alert mode.
- ▶ 4 preamble options to wake up properly the receiver.
- ▶ 2 motion detection options (enable / disable).
- ▶ 2 over temperature options (enable / disable).

In total, more than 300 000 combinations are possible within a given state machine to address the most demanding customer specification.

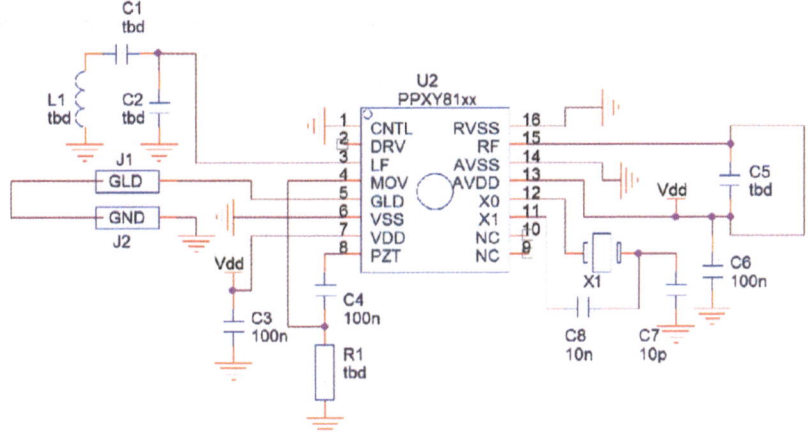

Fig. 7.   Application schematic example.

3.4 Application Example

Thanks to the high level of integration and extremely low power consumption, it is quick and easy and for the end user to build a cost effective wheel module (figure 7). The new sensor is delivered in a small outline 16 pin package and only a few external passive components, low frequency coil and crystal are required. Since there is no software development required, time to first wheel module prototype is significantly reduced.

The reduced number of external components also greatly enhanced the reliability of the wheel module by reducing the number of electrical connections and solder joints. The overall size and weight of the wheel module can be reduced to dimensions never before achieved.

Conclusion

Motorola's new dedicated Tire Pressure Monitoring Sensor is a unique opportunity for the automotive industry to implement a cost effective and reliable semiconductor solution inside the tire for continuous and accurate tire monitoring, thereby enhancing the overall safety of the car and its occupants. The combination of surface micromachined MEMS technology and CMOS ASIC is paving the way for future battery-less TPM solutions. Several battery-less architectures are under investigation by the industry (125 kHz transponder technology, UHF energize and back scatter modulation, solid-state batteries or even micromachined generators embedded in the tire). However, none of them can be efficiently implemented if the heart of the system consumes too much energy!

References

[1] Full text of the FMVSS138 is available at: http://www.nhtsa.dot.gov/cars/rules/rul-
     ings/TirePresFinal/index.html

[2] A 0.8 μm CMOS integrated surface micromachined capacitive pressure sensor
     with EEPROM trimming and digital output for a tire pressure monitoring system
     B.P. Gogoi, S. Jo, R. August, A. McNeil, M. Fuhrmann, J. Torres, T. F. Miller, A.
     Reodique, M. Shaw, K. Neumann, D. Hughes Jr. and D. J. Monk. Motorola
     Semiconductor Products Sector.

[3] Considerations to Improve Battery Life in Direct Tire Pressure Monitoring, SAE
     2002-01-1078, Mark L. Shaw, Motorola Semiconductor Products Sector US Patent
     6.472.243: Method of forming an integrated CMOS capacitive pressure sensor,
     Motorola Inc.

[4] Motion Sensing Techniques and Analysis for Direct Tire Pressure Monitoring, SAE
     2003-01-0202. Mark L. Shaw, Motorola Semiconductor Products Sector.

[5] US Patent 5.883.305: Tire Pressure Monitoring System, Motorola Inc.

[6] Suggested application notes
     ▶ AN1943/D: TPMS Demonstration kit, Jeff Burgess
     ▶ AN1951/D: Motorola Tire Pressure Monitor System Demo, Jeff Burgess
     ▶ AN1954/D MPXY8000 Series, Tire Pressure sensor, Ador Reodique
     ▶ AN1953/D: High Accuracy Digital Tire Pressure Gauge, Daniel Malik

Marc Osajda
Motorola Semiconductors
Sensor Products Division
Avenue Eisenhower
BP 1029
31023 Toulouse
France
marc.osajda@motorola.com

Keywords: MEMS, CMOS Pressure sensor, Tire Pressure Monitoring, State
              Machine

Low-Cost, Single-Chip Amplified Pressure Sensor in a Moulded Package for Tire Pressure Measurement and Motor Management

R. Bornefeld, W. Schreiber-Prillwitz, O. Stöver, Elmos Semiconductor AG
H. V. Allen, M. L. Dunbar, J. G. Markle, Silicon Microstructures, Inc.
I. van Dommelen, A. Nebeling, J. Raben, Eurasem B.V.

Abstract

Advanced automotive sensor systems used for tire pressure measurements and motor management show a strong demand for increased robustness, functionality and flexibility at reasonable low costs. To meat these requirements a single-chip sensor in a low cost package has significant advantages compared to other existing solutions. By combining silicon micro-machining designs and processes with advanced mixed-signal CMOS circuitry, many of the early limitations of single-chip sensors can be overcome. The result is a high-performance single chip pressure sensor that provides an amplified and calibrated output. Co-integration of sensor, CMOS amplification and EEPROM on a single chip results in a sensor structure that can be completely moulded into the package prior to calibration and compensation. The calibration and compensation parameters are then programmed into the package on an automated tester, providing a very low-cost high-volume sensor element.

1   Background

Piezoresistive pressure sensors need signal conditioning circuitry in order to be interchangeable and compatible with most electronic control systems. Each sensor element has individual characteristics and is sensitive to both pressure and temperature. The signal conditioning functions include calibration for offset and full scale variations to make each unit electrically interchangeable, temperature compensation for offset and full scale, and sometimes other functions as well, such as linearity correction, diagnostics, and filtering.

Traditional technologies to marry signal conditioning electronics with sensing elements have resulted in technical and commercial compromises in product cost and performance.

Parameter	Two-Chip Thick-Film Laser	Two-Chip Thin-Film Laser	One-Chip Thin-film Laser	One-Chip Analogue Correction with EEPROM	One-Chip Digital Correction with EEPROM
Circuit Complexity	Low and Low	Low and Moderate	Moderate to High	High	High
Process Complexity	Low	Moderate	Moderate to High	Moderate to High	Moderate to High
Component Costs	Highest	High	Low	Low	Low
Test Complexity	High	High	High	Low	Low
Test Cost	Very High	High	High	Low	Low
Package Size	Large	Good	Small	Small	Small
Package Cost	Very High	High	Low	Low	Low
Pressure Range Adaptability	Good	Good	Moderate	Moderate	Moderate
Trim after gel and/or packaging	No	No	No	Yes	Yes
Ability to do multi-order correction	No	No	No	Yes, with difficulty	Yes
Accuracy	Low	High	Moderate	Moderate	Very High
Resolution	High	High	High	High	Moderately High

Tab. 1. Comparison of various signal conditioning and error correction approaches.

Monolithic approaches have resulted in low cost and small size but in many cases have compromised device performance due to the restrictions place on circuit and sensor design by the monolithic fabrication process. Hybrid approaches use either a dedicated ASIC for signal conditioning or a discrete circuit approach. In many cases, these approaches have better performance than monolithic devices and offer flexibility for adapting new designs with simple components changes. However, these hybrid approaches are generally not as low in cost as the monolithic approaches and are also larger and. require additional assembly steps. More electrical and mechanical connections means more concerns about reliability. A comparison of different approaches is shown in table 1.

2   Applications

There are a number of applications such as Manifold Absolute Pressure (MAP) sensing and Tire Pressure Monitoring (TPM) that are driving the demand for integration as a way to lower system costs. In these automotive applications, especially, the need is for an integrated system design of sensor, signal processing and packaging together in order to realize the optimal solution.

There are a number of technologies to store the calibration and compensation values for each sensor ranging from thick film resistors that are laser trimmed after device testing to achieve temperature compensation and calibration to electronic trimming where the compensation and calibration values are electronically programmed into the chip – either with a one-time programmable technology such as fuses, zener diodes, on-chip trimmable resistors, or a re-programmable technology such as EEPROM.

Re-programmable technologies such as EEPROM allow coefficients to be programmed into the device multiple times, allowing for real-time programming during manufacturing or programming after assembly based on the test data for a given device.

3   Co-Integrated Approach

The co-integrated pressure sensor was collaboration between MEMS sensor designers and mixed-signal IC designers. This collaboration brings the strengths of each technology to the device design to maximize performance. The circuitry performs calibration for offset and sensitivity, and multi-order temperature compensation and error correction. On-chip EEPROM stores the calibration coefficients for the device. The manufacturing process flow has been designed to allow all standard CMOS steps to occur at the front-end of the process, with the micro-machining steps performed after the circuitry is completed.

The result shown in figure 1 is based on a standard 0.65 µm mixed-signal CMOS process with EEPROM. All of the CMOS processing is performed at the start of the process and the MEMS processing is performed at the end. This process flow results in fewer compromises on the circuitry and better sensor performance. Current chip size is 10 mm^2.

In addition, this "CMOS first" choice results in a better manufacturing flow because the CMOS process flow remains uninterrupted. With this approach, process controls, based on high volume production, are maintained. Sensor design has incorporated readily available standard P-type implanted sensing elements, eliminating the need to vary the process to incorporate an extra, special implant.

The MEMS process that follows the CMOS process steps does not adversely affect the CMOS functions. These processes include the silicon etch to form the pressure sensitive diaphragm, anodic bonding of a glass substrate for absolute pressure configurations, probing, wafer dicing, and inspection.

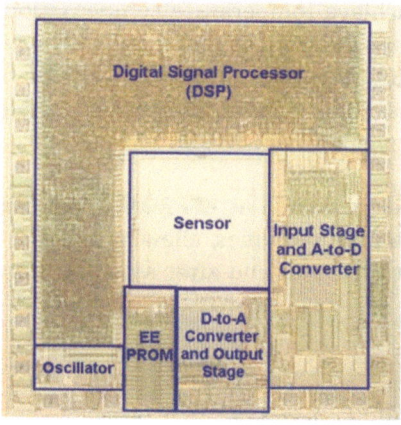

Fig. 1.   Photomicrograph of co-integrated pressure sensor.

4   Technical Details

The circuit on the co-integrated pressure sensor provides for a number of adjustable parameters. Based on production test data, the following parameters can be adjusted in the ASIC by programming the device using the on-board EEPROM.

▶ Pre-amplifier Gain – 32 to 152 in 16 steps (16 to 76 at 2.5 V offset).
▶ Pre-amplifier Gain sign – 1 bit (positive vs. negative output).
▶ Pre-amplifier Offset – 0 to 49.5 mV in 16 steps.
▶ Pre-amplifier Offset sign – 1 bit (positive vs. negative input).
▶ Upwards of 20 coefficients in Pressure and Temperature correction.
▶ Output Clamp – Independent Adjustment in 40 mV steps.
▶ Output Filter – 1 to 256 Conversions in 1, 2, 4, 8, 16, 32, 64, 128 and 256 steps equivalent to approximately 1 to 256 mS.

4.1 Circuit Block Diagram

The circuitry on the co-integrated pressure sensor includes all of functions for signal conditioning and calibration including amplification, correction, span calibration, temperature compensation, and multi-order non-linearity correction for pressure and for the temperature coefficients. A block diagram of the circuit is shown in figure 2. The system uses an 11-bit A/D converter with 8x over-sampling to provide for an effective 14-bit data conversion resolution. Data is then processed thru an on-board Digital Signal Processor (DSP) where the A/D data is corrected, based on stored calibration coefficients stored in the on-board EEPROM. The corrected signal is then fed to a 12-bit D/A converter, which in turn drives the output amplifier. The amplifier has been designed to be able to drive >2 nF of capacitance as needed for EMF suppression in the automotive environment.

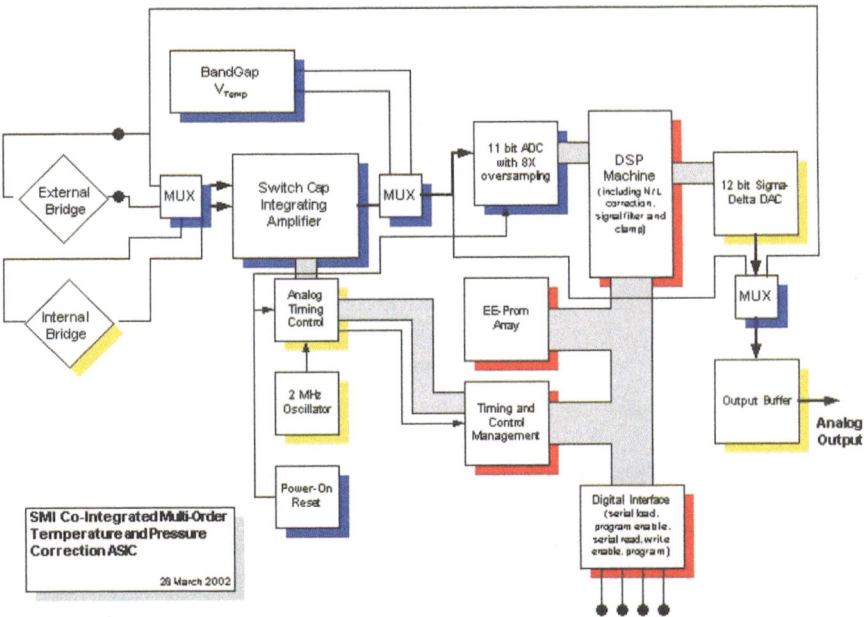

Fig. 2. Block diagram of circuitry on co-integrated pressure sensor.

4.2 Calibration

Calibration is done by measuring the sensor at multiple temperatures and pressures. Uncorrected data from the A/D converter is read out thru a digital I/O for each data point. The external calibration computer then determines the minimum error and loads the order of the correction in pressure and temperature and correction coefficients into EEPROM. Verification of the calibration can then be performed to verify calibration accuracy as desired.

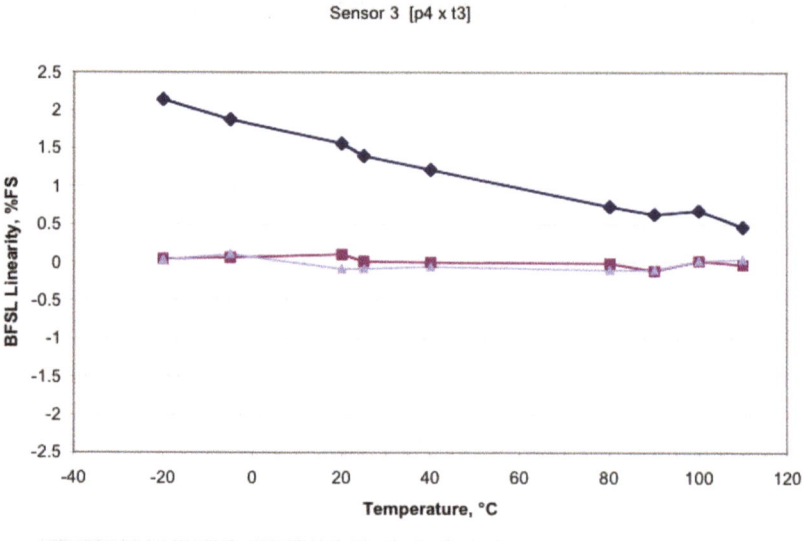

Best Fit Straight Line (BFSL) Linearity
M777.01B with 015-A Sensor
Sensor 3 [p4 x t3]

Fig. 3.   Linearity correction capability of co-integrated sensor system.

Figures 3, 4, and 5 show data from sensors before and after correction by the circuitry. The data in figure 4 shows the ability of the signal processor to compensate for pressure non-linearity, over at temperature range of -20°C to +110°C.

As shown in figure 4, the signal processor can also compensate to better than 0.2% for temperature coefficient of zero error, even when the sensor exhibits a 2nd order non-linearity in the temperature coefficient of zero.

Figure 5 represents the corrected temperature coefficient of span. The -20% change with increasing temperature of the uncompensated part gets corrected to close to zero. Total error over the compensated range for this sensor-system is less than 0.25%, including all errors combined including initial zero and full-scale gain-set errors.

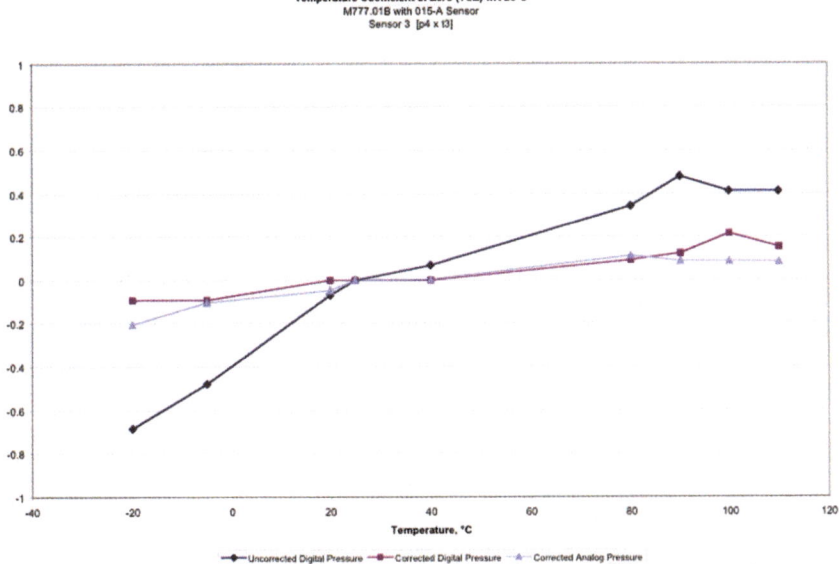

Fig. 4. Temperature coefficient of zero over temperature.

Over temperature, the performance of the device can change and thus calibration at other temperatures means increased accuracy at those temperatures. Examples of the ability of the DSP to correct errors in the sensor's output are shown in figures 3, 4 and 5. As can be seen in the figures, the algorithm used can be adapted to easily correct pressure non-linearities of various orders and further can facilitate correction that might be introduced by temperature dependent pressure non-linearities from gelling or other media interfaces.

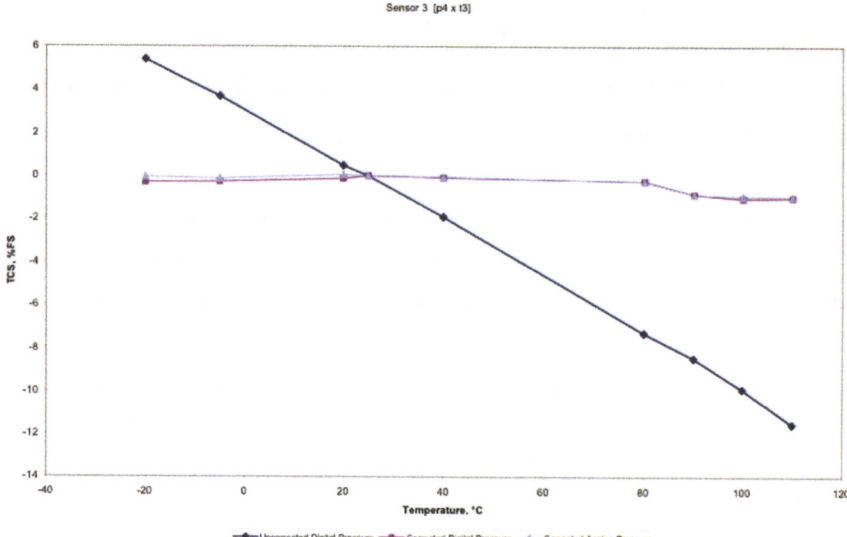

Fig. 5.   Temperature coefficient of span over temperature.

4.3   Sensor Construction

After completion of the CMOS processing steps, the wafers are shipped from the CMOS fab to the MEMS fab. The silicon etching is performed along with the final process steps for wafer bonding, dicing, and testing. The resulting die, a cutaway-cross-section of which is shown in figure 6, are then either shipped to die customers or assembled into a variety of packages.

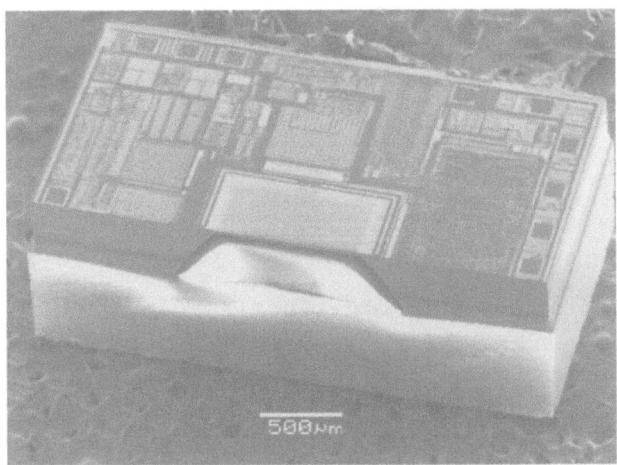

Fig. 6.   SEM cross-section showing CMOS and sensor area and bottom glass wafer.

4.4 Improvement over Two-Chip Solutions

By incorporating the DSP-based correction engine capable of multi-order correction for pressure and temperature errors, much better accuracy can be achieved with the co-integrated sensor than with a two-chip solution. For example, many two-chip solutions have good accuracy over temperature ranges near room temperature, but then have increasing error as temperature reaches the temperature extremes. With the co-integrated sensor, the error curves can remain the same over the entire temperature range.

4.5 Packaging

Co-integrated piezoresistive pressure sensors with digital correction engines and EEPROM calibration data storage have been reported before [1]. But up to now cost efficient packaging in the most used IC packaging technology plastic overmoulding failed due to different problems: packaging induced stress, thermal drifts and overstress caused by the moulding pressure [2]. By using a patented injection moulding approach [3] the packaging for the presented pressure sensor system is done in a 16-pin SOIC cavity overmoulded package. The special technique allows to expose the diaphragm area of the sensors chip as shown in figure 7, the wire bonds are overmoulded as in a standard IC package with the same level of proven reliability. Any stresses from die attachment, wire bonding, and moulding are compensated by programming the trim parameters in the final package. The result is a high performance part. This basic package is a foundation for other configurations, including further gel filling of the cavity. Various ports may also be welded on the package.

Fig. 7.   Single chip pressure sensor in SOIC 16 plastic package.

After programming, the EEPROM is electronically locked as a completed module, or further trimmed after system assembly. Because of the use of EEPROM, the trim parameters can be changed after initial programming, meaning that the trim values can be optimized during test or trimmed again after assembly of the sensor package into the system (for example, for serialization information or fine offset correction).

The resultant system essentially combines three technologies together to provide the user with a cost-effective solution:
- ▶ ASIC signal processing – ELMOS Semiconductor AG.
- ▶ Silicon micro machining and Test – Silicon Microstructures, Inc.
- ▶ Specialized Plastic Lead-frame packaging – Eurasem b.v.

5   Conclusion

A new, single-chip pressure sensor has been built with all signal conditioning and pressure sensing functions on the same chip. This co-integrated pressure sensor overcomes many of the performance limitations of previous single chip designs.

The co-integrated pressure sensor offers improved performance over previous single chip pressure sensor solutions while reducing size and cost compared to hybrid configurations. A process flow that is compatible with a standard 0.65 micron CMOS process has been used along with MEMS processing to create the pressure sensor with EEPROM on-chip to store the calibration and compensation values for the sensor. Test results show a dramatic improvement in device performance over the uncompensated result.

This co-integration technology can be extended beyond pressure sensors to other devices as well, including accelerometers, gyros, and other sensor types.

6   Acknowledgements

This sensor-system is a result of close coupling of the Design and Processing Groups at Silicon Microstructures, Inc; the Design and Processing Groups at ELMOS; and the Design and Processing Groups at Eurasem. The authors would like to acknowledge the contribution of each member of the team that made this device possible.

References

[1]   J.P. Schuster, W Czarnocki, X. Ding, B. Roeckner: Monolithic Pressure Sensor System with Digital Signal Processing, AMAA 99, Springer Verlag Berlin Heidelberg New York, pp297-309

[2]   J. Hesse, J.W. Gardner, W. Göpel: Sensors for Automotive Technology, Wiley-VCH Verlag 2003, pp195-196

[3]   I. van Dommelen: Plastic Packaging for Various Sensor Applications in the Automotive Industry, AMAA 2002, Springer Verlag Berlin Heidelberg New York, pp 289-296

Ralf Bornefeld, Wolfgang Schreiber-Prillwitz, Olaf Stöver
Elmos Semiconductor AG
Heinrich-Hertz-Str. 1
44227 Dortmund
Germany
rbornefeld@elmos.de

Henry V. Allen, Michael L. Dunbar, Jeffery G. Markle
Silicon Microstructures, Inc
1701 McCarthy Blvd.
Milpitas CA 95035
USA

Ignas van Dommelen, Andreas Nebeling, Jürgen Raben
Eurasem B.V
Microweg 1
6500 AN Nijmegen
The Netherlands

Keywords: co-integrated pressure sensor, digital calibration, plastic moulded housing, tire pressure measurement, motor management

Rollover Detection Systems

J. Haun, Delphi Electronics and Safety

Abstract

Rollover accidents often result in heavily injured or even killed passenger. Most often head injuries occur as a consequence of heads contact with the roof of the vehicle or the road when the head penetrates through the side window during rollover. Therefore industry is focusing more and more in rollover detection and rollover systems. In this paper new concepts and new algorithm design for vehicle rollover detection systems are presented. Delphi developed a fully tested system, which allows detecting rollover scenarios to trigger restraint systems as rollbars, belt pretensions, window airbags and/or head airbags.

1 Rollover Risks and Statistical Background

Rollover accidents are very heavy accidents. But different statistics are existing about rollover accident analysis. One reason for statistical variation is that these accidents are not counted as separate accidents. The police does not have separate sheets for statistical rollover analysis. As a consequence the number of cars in the statistic of a country is sometimes very low. The other reason is that a rollover usually does not occur alone. Most often a crash is followed by a rollover or vice versa. If one focuses on statistics with high number of accidents in different countries you can find also differences due to different vehicle population in a country and different street styles. For example in the United States you can find more SUVs (Sport Utility Vehicle) than in Europe or Asia. SUVs more often than other vehicles tend to roll because of the high centre of mass. Also in Eastern part of Germany there are more avenues than elsewhere. In case of a crash a car often hits a tree and sometimes tends to roll after this collision. In Great Britain a lot of detailed analyses about rollover accidents have been performed. The statistical data from British Police is very extensive and detailed. So this database can be used as a standard for Europe.

1.1 British Accident Data

British national data are commonly called "STATS 19" [1] due to the name of the form that the Police officers fill in for every road traffic accident involving an injury on a public highway. In this statistic about 200.000 accidents are collected per year. There are three different types of records: accident, vehicle and casualty records. An accident record is completed for each accident. A vehicle record is completed for every vehicle involved in the accident (even if no person was injured). The casualty record is completed for every injured person in the accident. The overall criteria for an accident to be included in the STATS 19 records are that a person must have been injured in an accident on a public highway. The data are available approximately 14 month after the year of collection in a form ready for electronic analysis.

Fig. 1.   Rollover accident in USA.

In figure 2 it can be seen that 19.3% of all Killed or Seriously Injured (KSI) persons are involved in rollover. This is even more than the rate without any element of roll (e.g. two cars crash front - front). So the highest rate of KSI accidents are rollover accidents. In 1999 in GB 2320 people died or have been seriously injured in cars with element of roll. In the British database a lot of very detailed information can be found (e.g. population characteristics, circumstances of accident, loading and road type as well as environmental conditions at the time of the accident.

1.2 US Accident Data

From 1999 to 2001 in USA the number of passenger vehicle occupants killed in all motor vehicles crashes increased by 4%, while fatalities in rollover crashes increased by 10%. In the same decade, passenger car occupant fatalities in rollover declined by 15 % while rollover fatalities in light trucks increased by 43%.

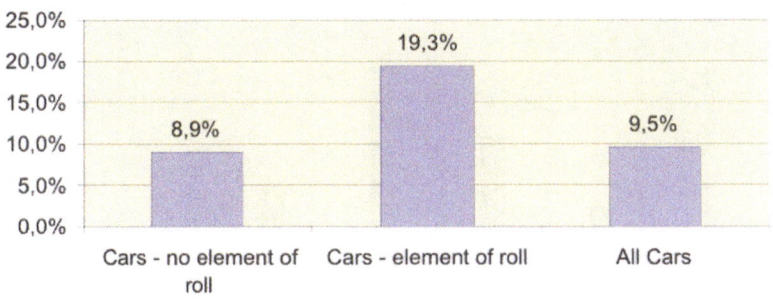

Percentage of Casualties With Fatal or Serious Injuries (KSI-Rate)

Fig. 2.   Percentage of casualties with fatal or serious injuries [1].

In the United States of America the U.S. Department of Transportation publishes an annual report [2]. In this document about 6.3 million accidents are reported per year. About 2 million of them are KSI types. Figure 3 shows one of the graphs out of the annual report of 2001. It can be seen that the risk of a rollover for pickups and light trucks SUVs is much greater than for vehicle cars. 19% of all fatal crashes are with element of roll. More details can be found in the report "Characteristics of Fatal Rollover Crashes" [3]. In U.S. rollovers account for 20% of passenger vehicles in fatal crashes. This is almost the same than in Great Britain. Due to this high amount of KSI accidents occur in combination with rollover (especially with SUVs) there are discussions about rollover safety systems for new vehicles. Especially for SUVs there is a lot of discussion going on.

To be able to compare useful rollover safety strategies it is important to define rollover resistance measuring methods. Therefore the NHFSA has done a so-called "New Car Assessment Program". The goal of this program was to find rules for the customer to assess the resistance of a car against rollover. In [4] this report can be found. NHTSA proposed to combine a vehicles Static Stability Factor (SSF) measurement with its performance in the so-called "Fishhook" manoeuvre. The Static Stability Factor of a vehicle is defined as the ratio of one half its track width to its center of gravity height (see www.nhtsa.dot.gov/hot/rollover/ for more ratings and explanatory information). The predicted rollover rate will be translated into a five-star rating system that is the same as the one now in use: One star is given for a rollover rate greater than 40%; two stars for rates between 30 and 39%; three stars for rates between 20 and 29%; four stars for rates between 10 and 19% and five stars are given for 10% or less.

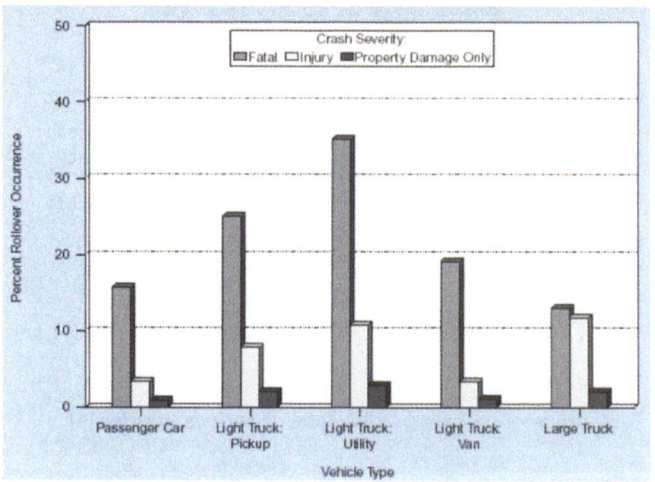

Fig. 3.   Percent rollover occurrence by vehicle type and crash severity [2].

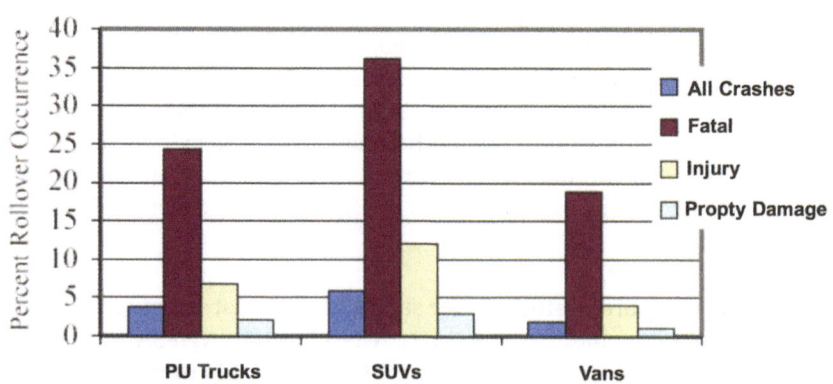

Fig. 4.   U.S. light truck rollovers by crash severity.
Source: NCSA, NHTSA, FARS 2000 and NASS/GES 2000

2   Rollover Sensors

The main sensor type that is used in rollover detection systems is the Angular Rate Sensor (ARS). These sensors can measure a roll rate, which is defined as the change of the angle by time. They are used also in space to measure the rotation of satellites or space shuttles and in aeroplanes to help the autopilot to work properly. Depending on the precision that is needed different physical effects are used in these sensors. In the following the main physical effects are explained.

2.1   Sagnac Effect

This effect is the most precise, but also most expensive method to measure roll rates. It is based on an optical phase shift of light. A laser beam is splitted in two beams and coupled anti-parallel into an optical gals fibre wire or interferometer. The fibre is winded up in many loops. The phase shift of the beams that are combined again after passing the fibre is a function of the roll rate of the fibre. In 1913 the French physicist Sagnac did a first experiment with some mirrors (figure 5).

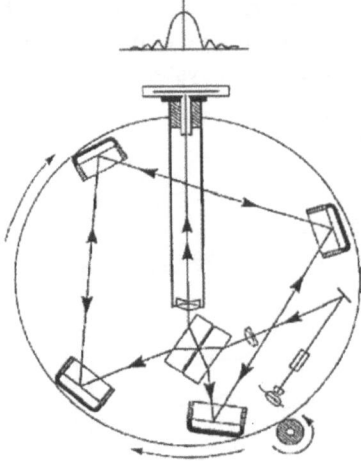

Fig. 5.   Principle of Sagnac effect [5].

In his experiment the light is splitted by a beam splitter. One part is mirrored right hand, the other left hand on a circle and combined again. When the whole experiment is rotated the interference pattern is changing. The relativistic interpretation of this phenomenon is that the light is passing the circle against the direction of rotation, which is moving, in a higher accelerated system. Therefore it is younger than the light that is passing the circle in the direction of rotation. Therefore the phase is different. The Sagnac phase can be written as:

$$\Delta\varphi = 8\pi \frac{A}{\lambda\, c_0} \Omega$$

Modern laser-optical rate sensors using this principle can measure rates in the range 0.0003°/s to 400°/s. The drift is less than 1°/h. These sensors are quite expensive and are used in the automotive industry only as reference sensors in crash tests. Sometimes the sensors are more expensive than the whole car.

2.2 Precision Effect

These types of sensors use the precision effect acting on mechanical gyroscopes (figure 6). A micro mechanical gyroscope is rotation with the angle velocity Ω. If the whole system is rotated with the angle velocity Ω, a torque M is occurring. As a result a force F can be measured. The Force F is a measure of Ω. The precision of these types of sensors is much lower than the optical, but they are cheaper. For automotive applications they are still too expensive, because of the micro-mechanical system that is needed.

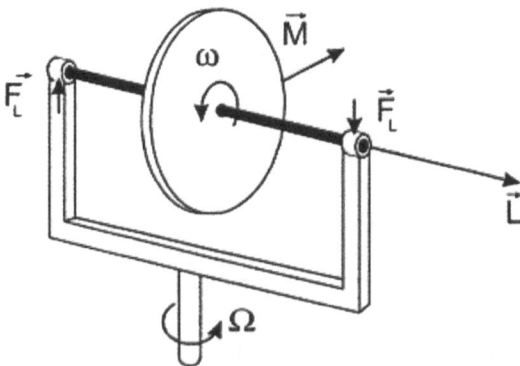

Fig. 6.   Precision effect acting on a gyroscope.

2.3   Coriolis Effect

The Coriolis force can be observed in non-inertial systems. It is called also a "virtual force". The principle can be seen in figure 7. A piece of mass is rotating on a circle at the angle velocity Ω. If the radius of the mass is changed, the inertia J is also changed. As a consequence a moment M is generated:

$$\vec{M}_C = \frac{\partial \vec{L}}{\partial t} = \vec{\Omega}\frac{\partial \vec{J}}{\partial t} = \vec{\Omega} \cdot m \frac{\partial}{\partial t}\vec{r}^2 = -2 \cdot m \cdot \vec{\Omega} \times \vec{v} \cdot \vec{r}$$

The Coriolis Force is given by:

$$\vec{F}_C = -2m\vec{\omega} \times \vec{v}$$

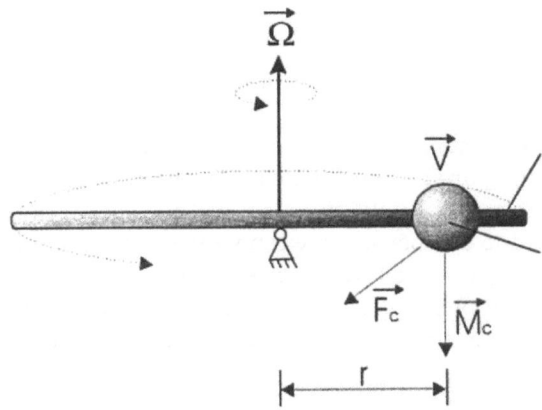

Fig. 7.   Coriolis force.

In sensors the piece of mass is a tuning fork (figure 8). The upper parts of this fork are oscillating. Therefore the radius r is changing permanently. If the whole tuning fork is rotated with at angular velocity Ω, the moment M is causing a change in the direction of oscillation. This additional motion is perpendicular to the basic oscillation and can be measured if the tuning fork is part of a capacitor, which changes the capacitance because of this additional motion. This is also a micro-mechanical system, but it is much cheaper than the gyroscope type. The price of such sensors is less than 50 Euro and they are used in automotive applications. They can measure roll rates up to 300%/s by a precision of about 1%/s. Drift of these sensors has to be compensated by software in the application. The primary oscillation is in the kHz-range.

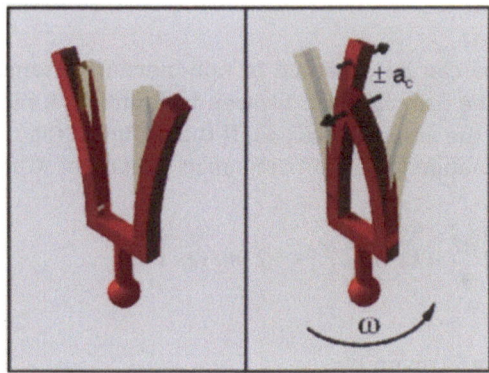

Fig. 8.   Tuning fork principle.

3   Rollover Detection

The main sensor in a detection system is the Angular Rate Sensor as written above. The signal of sensor is integrated by time to get the predicted roll angle. To be able to detect a non zero staring position (initial angle unequal 0°) additional sensors are required. For example a z-log-g sensor can measure the angle between the earth gravity field direction and the sensor position. The same sensor can be used to compensate a wrong angle calculation because of a possible bias. If the measured roll rate of the angular rate sensor is unequal 0° in case that there is no physical rate, the integrated angle would increase continuously. To avoid this a bias removal logic has to be implemented.

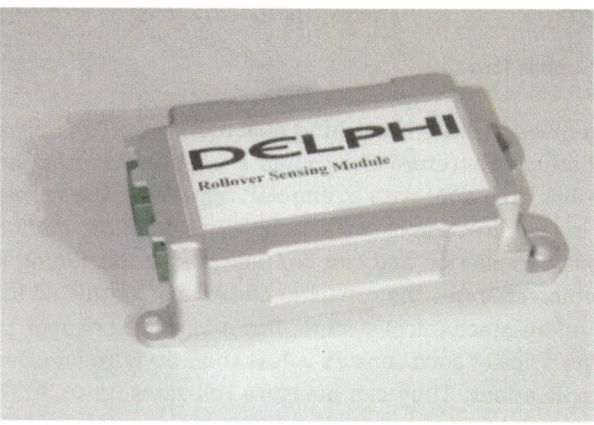

Fig. 9.   Delphi rollover sensing module.

It is also possible to measure the static roll angle by the use of perpendicular oriented linear (they measure in linear direction) accelerometers (y-low-g, x-low-g). In this case the Angular Rate Sensor can measure changes in the actual roll angle caused by high roll rates and the linear sensors can measure slow changes. Therefore a combination of such sensors result in the best-calculated angles. But it has to be considered that accelerations due to high speed curves, rough road conditions and misuse steering manoeuvres may occur. This can be achieved by useful software solutions.

4   Arming Concept

The entire measuring system consisting of sensors, micros and firing loops has to be verified by itself as good as possible. The reason is simple: It has to make sure that any failure in the system may not result in an unexpected deployment of the airbags or other restraint systems. The system is usually checked at power on and during normal operation. To improve the security level a secondary so called "arming path" is used. A deploy can occur only in case that both paths (main discrimination path and second arming path) detect a rollover situation. The arming path validates that an intended event is occurring. The goal is to reduce the number of unexpected deployments due to single-point electrical faults, unrecognised faults of the sensors and undesired trigger commands due to electro-magnetic induction.

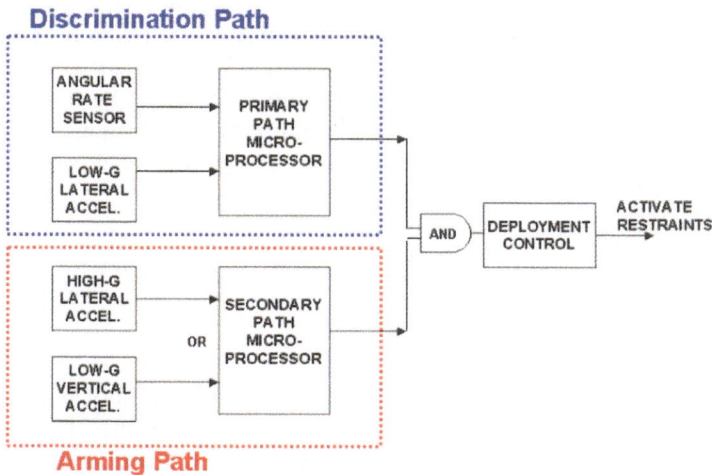

Fig. 10. Arming concept.

A simple (cheap) way of realisation can be used as arming path, but it should be as independent as possible. In other words the arming path should use other sensors, other microchips and a different software algorithm if possible. The bigger the independence is the bigger is the safety level. The concept is visualised in figure 10.

5   Calibration

Each rollover algorithm has to be calibrated. The calibration process is based on a set of rollover tests and misuse tests. The rollover tests should ensure that the algorithm detects the rollover situation early enough to deploy the restraint systems or the roll bars (in case of convertible cars). Normally several stages can be detected. For each stage (different roll rates or different maximal roll angles) there should be several rollover tests present for calibration. A set of misuse tests should ensure that there is no rollover detected even under rough driving conditions. Also airborne scenarios shall not result in an deployment if there is no rollover.

The calibration is tested in different ways. First on PC level a model of the whole rollover detection system is tested. All hardware including the tolerances and the software is included in this model. The calibration engineer can verify the behaviour of the system. Therefore he uses a library of real rollover, crash data and computer generated scenarios. Second the calibration is tested in the lab on the real hardware. Therefore the real module is modified on the bench. The electrical line between the sensor and the micro is cut and external electrical signals are connected to the input of the micro. Signals generated by the PC can simulate signals from the sensors. If the bench can work automatically, thousands of crash scenarios and rollover situations can be used to verify what the real hardware is doing. The output is also monitored and recorded by the PC.

6   Simulation and Modeling

Delphi developed several modern rollover concepts and is continuously improving them. To do this Delphi applies Physics, Math, Electronics and Software Engineering expertise to create auto-code rollover algorithm building blocks for next generation algorithms.

The software is tested automatically by special other software tools. By the help of this, software weaknesses can be found easily. For example it can be investigated how the software is reacting upon modification of the sensor signals as well as other hardware parameters or hardware tolerances. These investigations become more and more important, since it is not easy to find bugs in complex software.

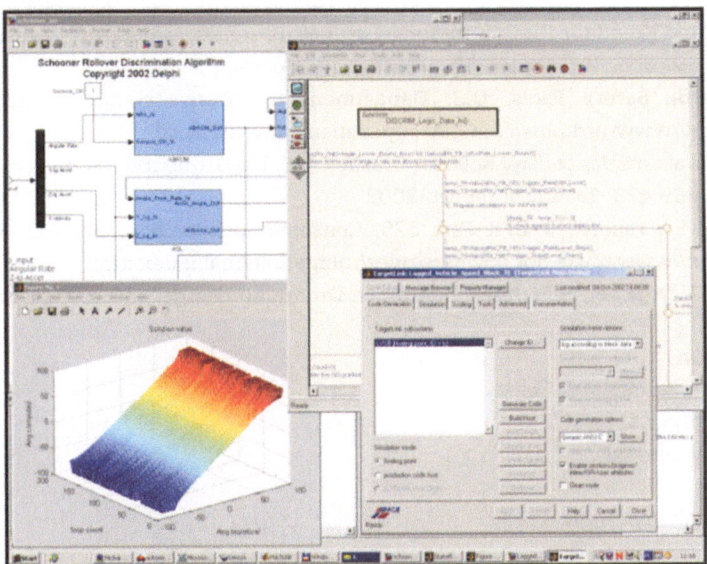

Fig. 11.   Simulation tools.

7   Outlook

Rollover has become more and more important in automotive industry. Especially for convertibles and SUVs this issue is discussed heavily. In the United States in the next years there will be some legal regulations expected. In the European Community a workshop including automotive suppliers, universities, OEMs and ensurance companies are working on rollover classification, modelling, sensing and useful safety strategies. It is possible to have standardised rollover tests as ENCAP tests on the market. Today a modern car should have a four or better five star rating in this test. Tomorrow the customers will also ask for the safety performance in a rollover condition. Even today you can find the first rollover crash tests in some automotive papers (e.g.

German ADAC). The knowledge and the technical realisation are present today. The customer's request will come tomorrow.

References

[1] STATS 19. Transport Statistics, Department for Transport, London, United Kingdom.1999 http://www.dft.gov.uk/stellent/groups/dft_transstats/documents/sectionhomepage/dft_transstats_page.hcsp

[2] Traffic Safety Facts, U.S. Department for Transportation, annual report, http://www-nrd.nhtsa.dot.gov/departments/nrd-30/ncsa/

[3] Characteristics of Fatal Rollover Crashes, U.S. Department for Transportation, DOT HS 809 438, Technical report, 2002

[4] NHTSA report 49 CFR Part 575, Consumer information, Rollover Resistance, http://www.nhtsa.dot.gov/cars/rules/ruling/RollFinal/index.html

[5] H. G. Deppner, STN Atlas Elektronik, Drehratenmessgeber, Bremen, 1999

Jens Haun
Delphi Electronics and Safety
Vorm Eichholz 1
42119 Wuppertal
Germany
jens.haun@delphi.com

Keywords: crash sensing, rollover detection, angular rate sensor, rollover, safety systems, NHTSA

Low-Cost Infrared Imaging Sensors for Automotive Applications

M. Hirota, Y. Nakajima, M. Saito, M. Uchiyama, Nissan Motor Co., Ltd.

Abstract

This paper describes two newly developed low-cost thermoelectric infrared imaging sensors, having a 48 x 32 element thermoelectric focal plane array (FPA) and a 120 x 90 element thermoelectric FPA, respectively, and two experimental vehicle systems. The FPAs provide high responsivity of 2.100 V/W and 3.900 V/W, respectively. The FPAs are basically fabricated with a conventional IC process and micromachining technologies and have a low cost potential. One experimental system incorporated in the Nissan ASV-2 is a blind spot pedestrian warning system that employs four infrared imaging sensors. This system helps alert the driver to the presence of a pedestrian in a blind spot by detecting the infrared radiation emitted from the person's body. The system can also prevent the vehicle from moving in the direction of the pedestrian. The other is a rearview camera system with an infrared detection function. This system consists of a visible camera and infrared sensors, and it helps alert the driver to the presence of a pedestrian in a rear blind spot. The sensor performance is suitable for use in various safety and occupant comfort systems.

1 Introduction

Advances in electronics in recent years have led to the adoption of many different types of sensors on vehicles, including pressure sensors, accelerometers and visible charge-coupled device (CCD) cameras. These devices are being used to meet societal demands for cleaner exhaust emissions, lower fuel consumption and improved safety and comfort, among other requirements. In Japan, the safety performance of vehicles has been a rising concern in conjunction with the increase in traffic accident fatalities since the late 1980s.

Infrared (IR) radiation refers to a form of light having wavelengths longer than those of visible light. Various types of warm bodies, including the human body, emit IR radiation or visible light corresponding to their temperature. Infrared sensors capable of detecting such radiated energy are an extremely useful means of human body detection because they can detect the presence of a person even at night without any illumination.

The IR radiation emitted by the human body, however, is of the long wavelength type (LWIR) in the 10 μm wavelength band, and its detection requires some ingenuity because of its extremely low energy level compared with that of visible light. Detection methods until the 1980s were limited to the use of cooled sensors (HgCdTe detectors, etc.) that were cooled to ultra-low temperatures in the vicinity of 77 K. Such sensors could not be used on ordinary passenger cars because of such aspects as the cooling system service life, their weight and power consumption.

However, two types of uncooled IR imaging sensor, which had been originally developed as a U.S. public sector technology [1,2], were implemented in commercial applications in 1992. These sensors provide detection performance nearly equal to that of cooled devices. The IR focal plane array (IR-FPA) used in these uncooled IR imaging sensors requires accurate temperature control in the vicinity of the phase transition temperature, necessitating the use of an additional device such as a Peltier thermoelectric cooler. Even uncooled IR imaging sensors, which have fewer components than cooled IR imagers, are still high in cost, a factor that limits their automotive use to high-end luxury cars at present. To promote widespread use of IR imaging sensors on vehicles in the coming years, it is necessary to reduce their cost further. We have focused on the thermoelectric type of IR-FPA [1-4] that is highly compatible with the conventional IC manufacturing process, does not require any temperature control mechanism or optical chopper and allows easy design of the second-stage processing circuit because of its thermoelectric nature, all of which help to give it a low cost potential. A CCD type of thermoelectric IR-FPA with over 10.000 pixels has already been announced [5]. However, we adopted a CMOS imager system [6,7] with the aim of reducing the cost further, and have attained high sensor performance by miniaturizing and optimizing the thermoelectric IR-FPA.

Moreover, we have used the results of our thermoelectric IR imaging sensor research to develop a prototype nighttime pedestrian warning system [8] and a blind spot pedestrian warning system. These systems were developed through our participation in the first and second phases of the Advanced Safety Vehicle (ASV) project promoted by the Ministry of Transport (currently the Ministry of Land, Infrastructure and Transport) beginning from 1991.

This paper describes the thermoelectric IR imaging sensors, the blind spot pedestrian warning system incorporated in the Nissan ASV-2 and a rear-view camera system with an IR detection function.

2   Infrared Imaging Sensor

2.1   Infrared Radiation Emitted by Human Body

Infrared radiation is a form of electromagnetic radiation having wavelengths from 0.78 to 1.000 μm and is one type of light wave. With respect to IR applications, the bandwidths having high transmittance in the atmosphere (atmospheric window) are divided into the short wavelength IR (SWIR) region, middle wavelength IR (MWIR) region and long wavelength IR (LWIR) region, as shown in figure 1. Selective use is made of these different wavelength regions depending on the target temperature.

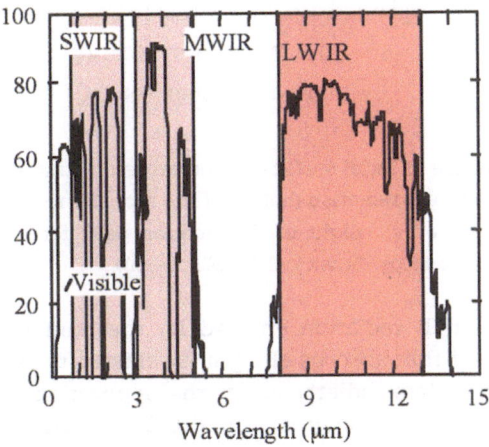

Fig. 1.   Spectral transmittance characteristics in the atmosphere.

The wavelength characteristics of the IR radiation emitted by the human body are determined by its radiant exitance, which is determined by the absolute temperature of the human body surface and its emissivity. As the first step of this study, the atmospheric temperature dependence of the facial surface temperature and the influence of solar radiation were measured in an environmental testing facility for automobiles. Under a shady condition, with only illumination and air-conditioning ventilation, the facial surface temperature showed a very small change of 4.1 °C in relation to an atmospheric temperature change of 30°C, as shown in figure 2. Even under a condition of mid-summer solar radiation (height of the illumination used for solar radiation was 3 m and the energy density at the floor was 2.76 MJ/m²·h), it was found that the facial surface temperature rose only 1°C compared with that under the shady condition. At an atmospheric temperature of 27°C (300 K) or lower, it was observed that a temperature difference of 5°C or more was obtained with respect to the facial surface temperature.

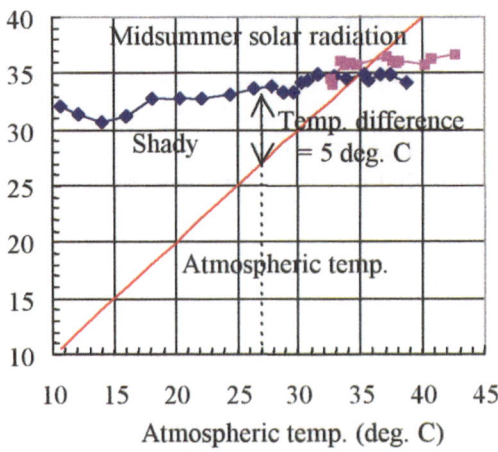

Fig. 2.   Dependence of facial surface temperature on atmospheric temper-ature and solar radiation (height of the face: 1.7 m; air-conditioning ventilation only; height of illumination used for solar radiation: 3 m; solar energy density: 2.76 MJ/m²·h).

The characteristics of IR radiation emitted by the human body were then determined. An object that absorbs all light is referred to as a blackbody, and it also becomes a complete radiator under the application of Kirchhoff's law. The emissivity of the human body is around 0.98, which is close to that of a blackbody. The spectral radiant exitance of a blackbody is determined by Planck's radiation law and is given by

$$W_\lambda = \frac{2\pi h c^2}{\lambda^5 \left(e^{hc/\lambda kT} - 1 \right)}$$

(1)

where W_λ is the emitted radiation per unit wavelength and unit area (W/cm²·μm), h is Planck's constant (= $6.63 \cdot 10^{-34}$ W·s²), k is the Boltzmann constant (= $1.38 \cdot 10^{-23}$ W·s/K) and T is the absolute temperature (K). The maximum radiation intensity wavelength λ_m can be found with Wien's displacement law as

$$\lambda_m T = 2897\mu\text{m} \cdot \text{K} \qquad\qquad (2)$$

As is clear from figure 2, the facial surface temperature is nearly constant in a range of 31-36°C (304-309 K). Therefore, the maximum radiation intensity wavelength of the IR radiation emitted by the human body is $\lambda_m \approx 9.5$ μm, as indicated in figure 3. Accordingly, use of LWIR radiation in a wavelength range of 8-13 μm is suitable for human body detection.

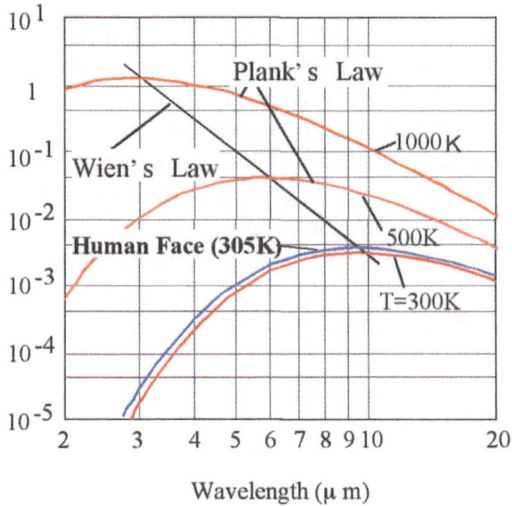

Fig. 3.   Spectral radiant exitance of a blackbody.

2.2 48 x 32 Element Thermoelectric IR Sensor

The thermoelectric IR-FPA consists of tiny thermocouples connected in series that utilize the Seebeck effect to convert the temperature difference between the hot and cold junctions into a voltage. In order to increase the temperature difference, the thermocouples are formed on a thin membrane that is fabricated in a micromachining process. The thermocouples are fabricated of p-type and n-type polysilicon layers, consisting of materials commonly used in general semiconductor processes and connected in series.

Figure 4 shows the functional principle of the sensor element. Incident IR radiation is absorbed by the Au-black absorber in the center of the sensor and converted to thermal energy, which is transmitted in turn via the hot junctions, beam and cold junctions by means of heat radiation to the Si substrate that serves as a heat sink. To increase the thermal resistance between the hot and cold junctions, a portion of the Si substrate is removed to form a cavity. In addition, high sensitivity is attained by using a vacuum-sealed package to avoid internal heat conduction through the air. The responsivity R of the thermoelectric IR-FPA is given by

$$R = n \cdot \alpha \cdot R_{th} \cdot \eta \tag{3}$$

where n is the number of thermopile pairs, a is the Seebeck coefficient (sum of p-type and n-type), R_{th} is the thermal resistance between the hot and cold junctions and η is infrared absorptivity. Equation (3) indicates that increasing η and R_{th} is effective in improving responsivity R. The optimum values of η and R_{th} were calculated on the basis of the sensor size and the accuracy of the fabrication process. Additionally, a precisely patterned, high-IR absorptivity Au-black layer, which was developed independently in our laboratories, was adopted to improve η. As a result, absorptivity of more than 90% has been achieved around a wavelength of 10 µm. A scanning electron microscope (SEM) micrograph of one element is shown in figure 5.

This pixel structure was then used to fabricate a prototype 48 x 32 element thermoelectric IR-FPA. A micrograph of the 48 x 32 element IR-FPA chip is shown in figure 6. The dimensions of the various parts of the sensor are as follows: chip size of 10.5 x 7.44 mm², pixel pitch of 190 x 190 µm² and thermopile width of 0.8 µm for the six pairs of thermopiles. The positive terminal of each pixel is connected to an output line through two built-in NMOS transistors controlled by an external input signal. The negative terminal of all the pixels is connected to the same common terminal. Additionally, because this device has comparatively high internal resistance of 116 kW, it is necessary to limit the bandwidth of the second-stage circuit for the purpose of reducing Johnson noise (thermal noise). In order to obtain the desired video frame rate, the entire sensor is divided into 12 blocks of four columns each, and signals are output in parallel from 12 output lines. This vacuum-sealed IR-FPA achieves responsivity of R = 2,100 V/W and a thermal time constant of 25 ms, representing high levels of performance compared with similar commercial devices.

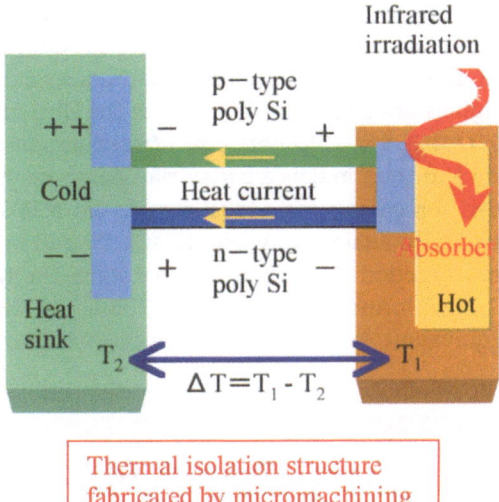

Fig. 4.   Functional principle of the sensor element.

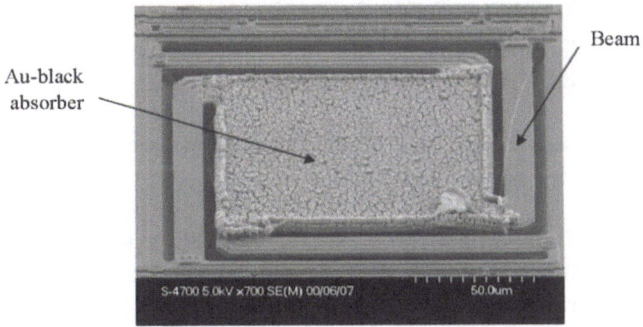

Fig. 5.   A SEM micrograph of one element of the 48 x 32 element thermoelectric IR-FPA.

This thermoelectric IR-FPA with its low cost potential was then incorporated
into an IR imaging sensor that has been developed independently in our laboratories for automotive applications. A photograph of the IR imaging sensor is
shown in figure 7, and its specifications are given in table 1. The IR imaging
sensor is a compact, lightweight device, measuring 100 mm in width, 60 mm
in height (excluding the bracket) and 80 mm in depth and weighing 400 g. The
sensor has a germanium meniscus single-element lens with a focal length of
f = 15 mm and f/0.7. Both sides of the lens are coated with an anti-reflection
coating having a wavelength of 10 μm. This IR imaging sensor operates under

a program stored in a Read Only Memory (ROM) incorporated in the Central Processing Unit (CPU, SH7034). The 2-D array sensor outputs successive signals in parallel from the 12 blocks based on an address signal request from the CPU. The output signals are amplified and sent to the CPU via an 8-bit analog-to-digital converter (ADC). The CPU performs offset compensation and responsivity compensation and then makes a pedestrian detection judgment. A function is also included for converting video data to an NTSC video signal by means of a video digital-to-analog converter (DAC). This IR imaging sensor can output a pedestrian detection signal and image data in the NTSC video signal format via the RS-232C interface. With its f/0.7 lens, the IR imaging sensor achieves a Noise Equivalent Temperature Deviation (NETD) of less than 0.4°C.

Fig. 6.   Micrograph of the 48 x 32 element thermoelectric IR-FPA (chip size: 10 x 7.44 mm^2).

Figure 8 shows a facial image of a person wearing eye glasses that was obtained with this IR imaging sensor. The eyeglasses portion appears as a dark region because infrared rays in the 10 μm wavelength band do not pass through glass.

Fig. 7.   Automotive IR imaging sensor incorporating the 48 x 32 element thermoelectric IR-FPA.

Performance Parameter	Capability (@300K)
Array configuration	48 x 32 elements
Chip dimensions	10.5 mm x 7.44 mm
Pixel size	190 μm
Spectral response	8 – 13 μm
Responsivity	2100 V/W
Thermal time constant	25 ms
Lens	f = 15 mm, f/0.7
NETD (@ f/0.7)	0.4°C
Field of view	30° x 20°
Outputs	RS-232C, Detection, NTSC
Power	12 V, 0.25 A
Overall dimensions	100 mm x 60 mm x 80 mm
Weight	400 g

Tab. 1.   Specifications of the IR imaging sensor.

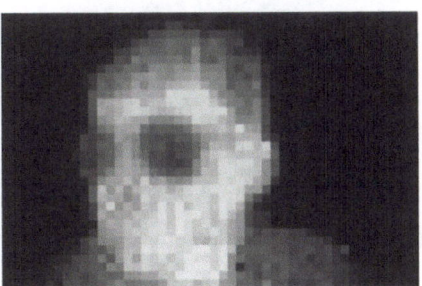

Fig. 8.   IR image obtained with the automotive IR imaging sensor.

2.3   120 x 90 Element Thermoelectric IR Sensor

A prototype 120 x 90 element FPA with a 100 μm element pitch was developed with the aim of increasing the resolution and detection range. The element pitch was narrowed from 190 μm to 100 μm to reduce the overall size of the chip. To maintain the desired performance despite the smaller element pitch, thermal resistance R_{th} was increased by reducing the beam width from 14 μm to 4.4 μm. A micrograph of the prototype FPA is shown in figure 9. The overall size of the chip is 14.4 mm x 11.0 mm. The imaging area is 12.0 mm x 9.0 mm because the element pitch is 100 μm. In order to reduce the scanning frequency and Johnson noise, the device is divided into twelve blocks (Block #1-Block #12) for every ten columns, similar to the 48 x 32 element FPA, and is scanned in parallel. In other words, this device operates as twelve 900-element FPAs arranged in parallel. Each element consists of two pairs of a p-n polysilicon thermocouple and an NMOS transistor for Y-address selection. The measured electrical resistance of the prototype sensor was 90 kΩ. An FPA was mounted in a vacuum-sealed package with a Ge window and responsivity was measured by using infrared radiation emitted from a blackbody furnace at 500 K. The average responsivity of all 10.800 elements, however, was 3.900 V/W, which agreed well with the calculated results and is higher than any previous value reported for thermoelectric FPAs. A prototype IR imaging sensor incorporating the 120 x 90 element thermoelectric IR-FPA is shown in figure 10. The IR imaging sensor consists of two units. A head unit equipped with the IR-FPA and a Ge lens measures 136 mm in width, 112 mm in height (excluding the bracket) and 35 mm in depth. The Ge lens is a meniscus single-element lens with a focal length of f = 15 mm and f/0.7. Infrared images taken with the FPA are shown in figure 11. These images are shown with offset and responsivity compensation.

Fig. 9.   Micrographs of a 120 x 90 element focal plane array.

Fig. 10.   A prototype IR imaging sensor incorporating the 120 x 90 element thermoelectric IR-FPA.

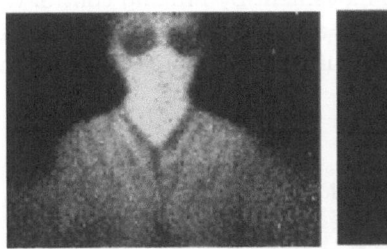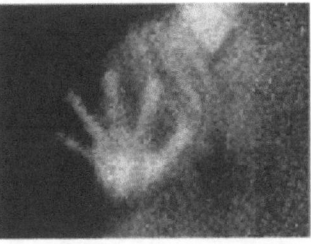

Fig. 11. Infrared images taken with a 120 x 90 element focal plane array

3 Blind Spot Pedestration Detection System Research

3.1 Concept

This system was incorporated in the Nissan ASV-2. The concept of this system is to detect infrared radiation emitted by the human body or other objects with the aim of helping to prevent blind spot accidents by alerting the driver to be careful and by preventing the vehicle from moving in the direction of a detected person. This is accomplished by using four IR imaging sensors, mounted at the front and rear, which detect children or other heat sources in the driver's blind spots (within 3 m and within the field of view of the sensors) where they cannot be seen and could become obstacles as the vehicle begins to move. The concept of the system is illustrated in figure 12.

Two new functions were added to this system, taking into account the characteristics of blind spot accidents. One is a human body detection function using four IR imaging sensors, and the other is a function for preventing the vehicle from moving by means of braking control. There can be a time difference ranging from several seconds to several minutes between the time a driver enters a vehicle and starts the engine and the time the vehicle is put in motion. It is possible that the circumstances around the vehicle may change during that interval. Because children in particular are apt to do unexpected things, incidents have been reported where blind spot accidents occurred even though the driver confirmed the safety of the environment around the vehicle before getting in. The IR imaging sensors, capable of detecting IR radiation emitted by the human body regardless of whether it is day or night, were adopted for this system because it was thought that a function for selectively detecting the human body and other heat sources would be the best way of helping to prevent such blind spot accidents.

The function for preventing vehicle movement by braking control was added because there are instances when the extremely short distance between a vehicle and an obstacle does not allow sufficient time for the driver to execute an evasive maneuver.

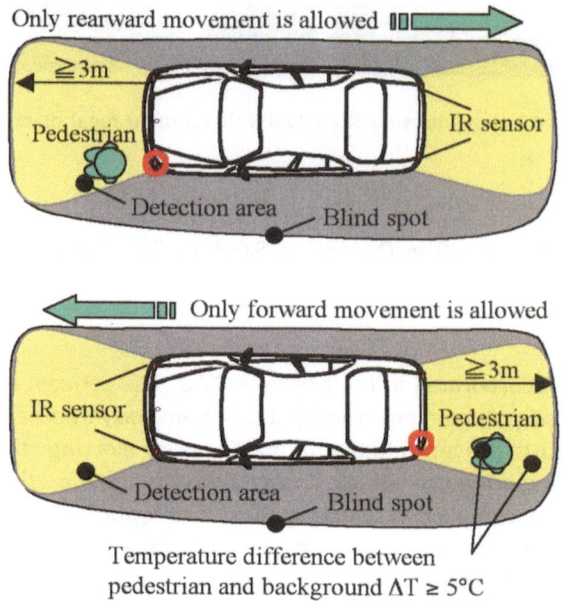

Fig. 12. Concept of the blind spot pedestrian warning system.

3.2 System Configuration

The system configuration is shown in figure 13. Infrared imaging sensors mounted at the vehicle's four corners serve to detect a pedestrian in the driver's blind spots. In the event a pedestrian is present in the direction in which the vehicle is about to move, the system issues audible and visual warnings to alert the driver and activates throttle and brake actuators to prevent the vehicle from moving. The components making up the system include: four IR imaging sensors incorporating a pedestrian detection judgment capability; a blind spot pedestrian warning control unit that integrates the signals from the IR imaging sensors and sends a signal to the Control Area Network (CAN); a vehicle motion control unit or an Auto Box (main controller) that manages vehicle motions; an integrated Human-Machine Interface (HMI) control unit; a monitor and a warning unit; and throttle and brake actuators that prevent the vehicle from starting off.

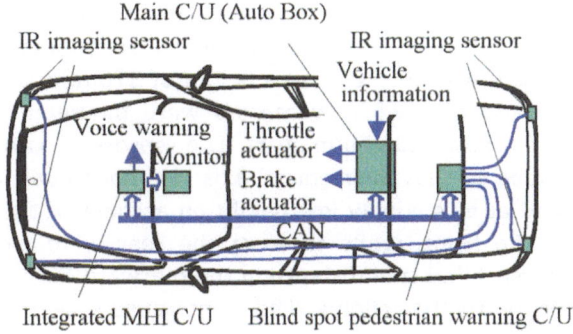

Fig. 13.   Overview of the blind spot pedestrian warning system.

3.3   Detection Range of IR Imaging Sensors and Mounting Positions

The mounting positions of the IR imaging sensors are closely related to their detection range. A study was made of the mounting positions under the following conditions.

▶ Two IR imaging sensors each would be used at the front and rear, for a total of four sensors.
▶ The sensors would have a field of view of 30°. (horizontal) by 20°. (vertical).
▶ The detection range would be 3 m under the condition that the temperature difference $\Delta T = T_t - T_e = 5°C$.
▶ The sensors would be mounted so that they did not interfere with the lamps, cooling equipment, tailpipe and other vehicle parts.

An experiment was conducted to measure the potential blind spot areas around the Nissan Cima (FY33) sedan that was the base vehicle of the Nissan ASV-2. The eye point in the driver's seat was set at a height of 1.05 m from the ground. Measurements were made of the areas where the ground (ground height of 0 m) could not be seen directly with the eye or by means of the mirrors from the driver's seat, assuming that the task was to detect a human body lying on the ground. The results are shown in figure 14. It was found that the blind spots at the front were within a distance of 2.7-4 m from the vehicle body and those at the rear were within a distance of 9-11 m. Taking those results into account, the positions of the four IR imaging sensors were determined so that they satisfied the four conditions mentioned above.

Photographs of the test vehicle are shown in figure 15. In figure 15b, two IR imaging sensors of the same type as those used at the front are mounted inside a germanium window at the rear. As indicated in figure 15a and 15b, the two front sensors were installed at a height of 370 mm and were spaced 1.580 mm apart, and the two rear sensors were mounted at a height of 515 mm and were spaced 1.000 mm apart. The sensors at both the front and rear were positioned so that their optical axis was horizontal. The front sensors were installed at a relatively low height on account of the shorter blind spot distance at the front, while those at the rear were mounted higher because of the longer blind spot distance in the rearward direction and also to avoid interference with the tailpipe. The front sensors were spaced farther apart because a larger detection range was needed at the front than at the rear in connection with the front-wheel steering of the vehicle. The front and rear detection ranges thus determined for the system are indicated in figure 14b. With the present system, there are small areas very close to the front and rear of the vehicle where detection is not provided. To cover these areas as well, either the field of view would have to be expanded or approximately three sensors each would have to be installed at both the front and rear.

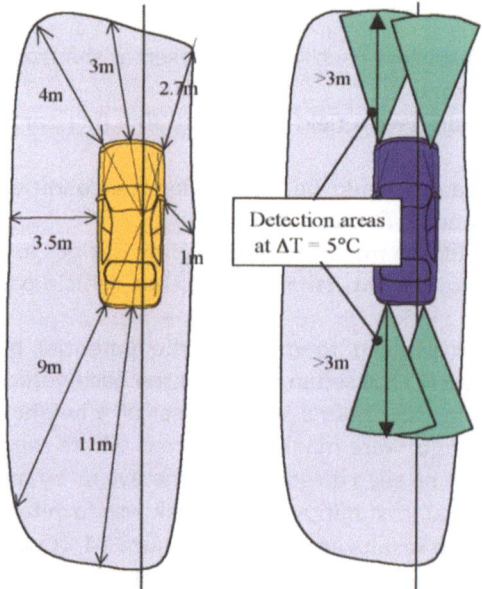

Fig. 14. Road surface areas that cannot be seen from the driver's seat of the Nissan ASV-2 base vehicle (Cima) at an eye point height of 1.05 m from the ground (left) and detection areas at $\Delta T = 5\,°C$ (right).

3.4 System Operation

The flowchart in figure 16 outlines the operational sequence of the system. At the time the vehicle is stopped, it is assumed that there are no pedestrians in the front and rear detection ranges. In response to a request signal from the Auto Box, the IR imaging sensors capture baseline image data (image A) and record the data in their internal memory units. When a vehicle is parked for a long time, it is presumed that the atmospheric temperature and sunlight will change, so the baseline video data (image A) are updated accordingly. The IR imaging sensors take images at periodic intervals of approximately 30 s and update the baseline image data (image A) if there is no heat source in the image. The baseline image data are not updated when a heat source is present.

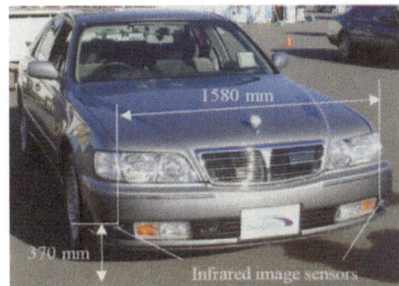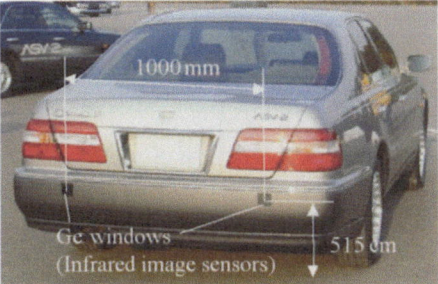

Fig. 15. Nissan ASV-2 equipped with the blind spot pedestrian warning system. (a) Front view, (b) rear view (with Ge window) and (c) enlarged front view.

Also, at the time the vehicle is about to start off, the IR imaging sensors capture start-off image data (image B) in response to a request signal sent from the Auto Box. To remove the influence of background heat sources, a differential image (image C) is calculated by subtracting image A from image B. The IR imaging sensors then process differential image C to detect heat sources. When a heat source is detected, a detection signal and the position of the sensor that detected the heat source (e.g. right front) are sent to the Auto Box via the blind spot pedestrian warning control unit. If the position of the sensor that detected the heat source (i.e. front or rear) coincides with the vehicle's intended direction of movement, the Auto Box judges that a pedestrian is present in that direction. It then alerts the driver by means of an audible warning and a display screen indicator, using signals sent through the integrated HMI control unit.

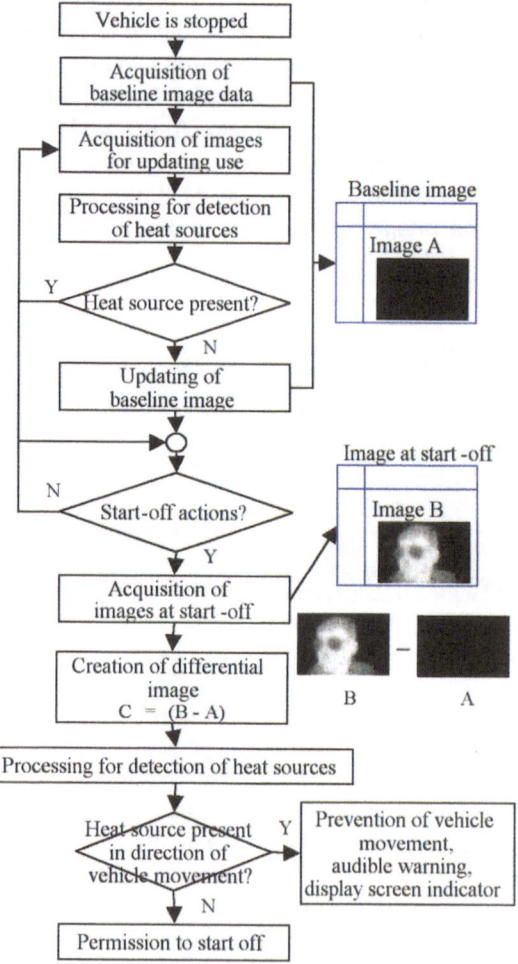

Fig. 16. Operational sequence of the blind spot pedestrian warning system.

The Auto Box also activates brake and throttle actuators that forcibly prevent the vehicle from moving in the direction of the detected pedestrian. Conversely, if the sensor position and direction of movement do not coincide, the Auto Box judges that there is no pedestrian in the vehicle's direction of movement and allows the vehicle to proceed.

The system judges that a heat source is present if an object corresponding to a temperature difference of 4°C or greater is in 10 or more pixels of the above-mentioned differential image C. That number of pixels is approximately equal to the size of a human face at a distance of 3 m (corresponding to a 15 cm angle).

3.5 Detection Results

Human body detection distances were measured under different atmospheric temperature conditions to verify the operation of the prototype system. The results are given in figure 17. In a low temperature range of 10-22°C, the detection distance was 6-8 m, which would cover a large portion of the rear-ward blind spot shown in figure 11. On the other hand, the detection distance decreased to 3-3.5 m in a temperature range of 24-30°C. Under a condition of a temperature difference $\Delta T = T_t$ (surface temperature of a pedestrian) - T_e (background temperature) >5°C (i.e., the atmospheric temperature (background temperature) is no more than 27°C), a detection distance of 3 m or greater was obtained with the prototype system specifications. Additionally, it was also confirmed in actual vehicle tests conducted with the Nissan ASV-2 that the system detected the presence of a human body and prevented the vehicle from moving in that direction. These results provide confirmation that the prototype system is effective in helping to protect pedestrians in potential blind spots.

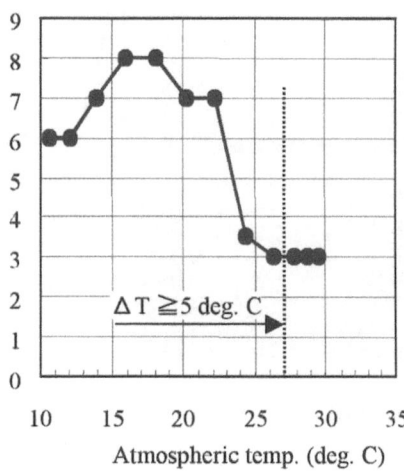

Fig. 17. Relationship between IR imager detection range and atmospheric temperature.

4 Rearview Camera with IR Imaging Sensors

This system is a fusion of a visible rear-view camera and the IR imaging sensors described above, enabling it to alert the driver to the presence of pedestrians in the rearward blind spot. The in-vehicle configuration of the system is shown in figure 18, and the positions of the devices are shown in figure 19.

An image processor combines the image signal from the visible camera with the detection results of two IR imaging sensors to produce a rearview image that is shown on a dashboard display screen. As shown in figure 18, the IR detection area achieved with the two IR imaging sensors is a trapezoid that measures 1.7 m along the width of the vehicle), 2.7 m on its long side and 2.1 m in distance from the vehicle. The long-side dimension with the IR imaging sensors and the distance along the ground are 2.7 m. A detected heat source is indicated on the screen by a red dot at the center of gravity of the detected region as shown in figure 20, and an audible warning is also given to alert the driver.

Based on the results shown in figure 17, the system is capable of detecting a human body in the entire detection area under a condition of $\Delta T = T_t - T_e > 5\,°C$.

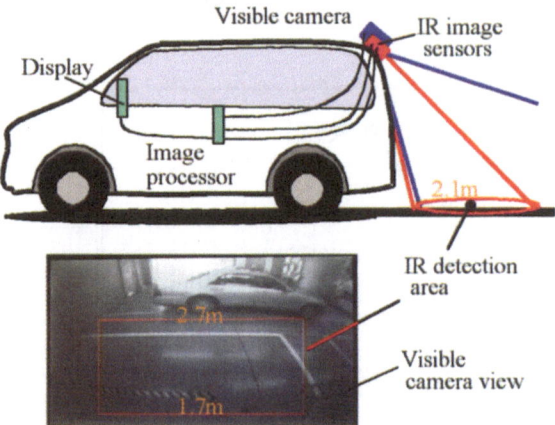

Fig.18.   System configuration.

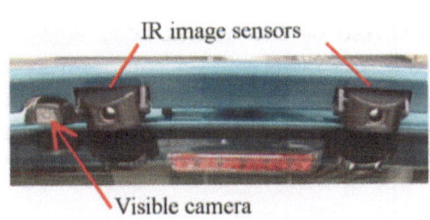 

Fig. 19. IR camera system installed on Nissan ASV-2. (a, left) Enlarged view of installation positions and (b, right) entire rear-end.

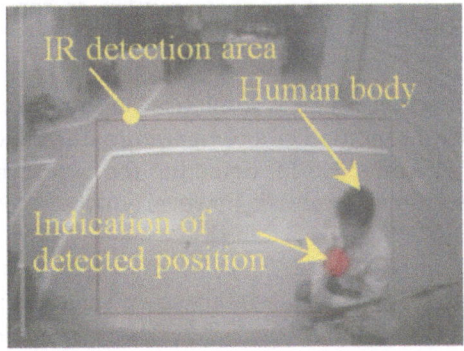

Fig. 20. Display screen showing detected human body.

5   Other IR Sensing Applications

Occupant detection systems for a smart air conditioner, an intruder alarm, a advanced air bag technology (AABT) have been developing. Prototype systems are equipped on the experimental vehicle illustrated in figure 18. Another system for recognizing hand commands is also being developed.

6   Summary

The characteristics of IR radiation emitted by the human body were investigated by measuring the dependence of the facial surface temperature on the atmospheric temperature and solar radiation. The results made it clear that the facial surface temperature shows little change under a condition of a light wind.

Thermoelectric IR imaging sensors with 48 x 32 elements and 120 x 90 elements, respectively, were developed that have a low cost potential. The 48 x 32 element sensor was used to develop a prototype blind spot pedestrian warning system that was incorporated in the Nissan ASV-2. In addition, it was also used to develop a prototype rearview camera system with IR detection capability.

The human body detection range is influenced by the background temperature and is 3 m under a condition of a temperature difference $\Delta T = 5\,^{\circ}\text{C}$. This detection range may be effective for detecting the presence of pedestrians in potential blind spots around a vehicle, however, further research work in this area is needed.

In future research work, it will be necessary to enhance the responsivity of the IR imaging sensor so that it functions more effectively under various environmental conditions and to improve the weatherability of the optical system.

References

[1]   C. Hanson, "Uncooled thermal imaging at Texas Instruments", SPIE, Vol. 1689, pp. 330-339, 1993.

[2]   R. A. Wood, C. J. Han, and P. W. Kruse, "Integrated uncooled IR detector imaging arrays", Proceedings of the IEEE Solid State Sensor and Actuator Workshop, pp. 132-135, 1992.

[3]   JJ.Yon, E. Mottin, L. Biancardini, L. Letellier, and JL. Tissot; "Infrared Microbolometer Sensors and Their Application in Automotive Safety", Advanced Microsystems for Automotive Applications 2003, Springer Verlag, ISBN 3-540-00597-8, pp 137-157.

[4]   T. Ishikawa, M. Ueno, K. Endo, Y. Nakaki, H. Hata, T. Sone, M. Kimata, T. Ozeki, "Low-cost 320 x 240 uncooled IRFPA using conventional silicon IC process", SPIE, Vol.3698, pp.556-564, 1999.

[5]   T. Kanno and M. Saga, "Uncooled infrared focal plane array having 128 x 128 thermopile detector elements", SPIE, Vol. 2269, 1994.

[6]   M. Hirota and S. Morita, "Infrared sensor with precisely patterned Au-black absorption layer", SPIE, Vol. 3436, pp. 623-634, 1998.

[7]   M. Hirota, Y. Nakajima, M. Saito, F. Satou, and M. Uchiyama, "120 x 90 element thermopile array fabricated with CMOS technology", SPIE, Vol. 4820, pp. 239-249, 2002.

[8]   M. Hirota, M. Saito, S. Morita, and H. Fukuhara, "Nighttime Pedestrian Monitoring System and Thermal Infrared Technology", Jidosha Gijutsu, Vol. 50, No. 11, pp. 58-63, November 1996 (in Japanese).

Masaki Hirota, Yasushi Nakajima, Masanori Saito, Makoto Uchiyama
Nissan Motor Co., Ltd.
1 Natsushima-cho, Yokosuka
Kanagawa 237-8523
Japan
m-hirota@mail.nissan.co.jp

Keywords: infrared, sensor, thermopile, automotive, blind spot, pedestrian detection, occupant detection

Application of Micro Scanning Laser Doppler Vibrometer for Vibrational Velocity and Resonance Frequency Measurement on Micro Electro Mechanical Devices (MEMS)

P. Castellini, B. Marchetti, E. P. Tomasini, Università Politecnica delle Marche

Abstract

Micro-Electro-Mechanical Systems (MEMS) represent the integration of mechanical elements, sensors, actuators, and electronics on a common silicon substrate through the utilization of microfabrication technology. The "micromachining" processes selectively etch away parts of the silicon wafer or add new structural layers to form the mechanical and electromechanical devices. MEMS promises to revolutionize nearly every product category by bringing together silicon-based microelectronics with micromachining technology, thereby, making possible the realization of complete systems-on-a-chip. MEMS is truly an enabling technology allowing the development of smart products by augmenting the computational ability of microelectronics with the perception and control capabilities of microsensors and microactuators. MEMS is also an extremely diverse and fertile technology, both in the applications it is expected to be used, as well as in how the devices are designed and manufactured. MEMS technology makes possible the integration of microelectronics with active perception and control functions, thereby, greatly expanding the design and application space.

Nevertheless, the spread of the micro-electro mechanical systems (MEMS) market is strictly related to the development of manufacturing technologies and to the implementation of analysis and diagnostic technique that allow to increase performances and reliability.

In this paper we demonstrate the feasibility of the laser doppler vibrometer, which has been designed and developed for measuring in the microscale, to characterize the dynamic of MEMS, in terms of measuring the vibration velocity in several point of the sample and revealing the resonance frequencies of the structure.

1 Introduction

In a previous work [2], the design and development of a workstation based on a laser Doppler vibrometer has been presented. In this paper we will demonstrate the application of such measurement technique on MEMS devices in order to study their vibrational behaviour

Microelectromecanical systems (MEMS) refer to devices that have a characteristic length of less than 1 mm but more than 1 μm, that combine electrical and mechanical components and that are fabricated using integrated circuits batch-processing technologies.

MEMS are finding increased applications in a variety of industrial and medical fields. The automotive Industry uses MEMS pressure sensors to measure engine oil pressure, vacuum pressure, fuel injection pressure, transmission fluid pressure, ABS line pressure, tire pressure, and stored air bag pressures. MEMS accelerometers can be used to trigger the air bag or lock seat belts. MEMS temperature sensors can be used to monitor oil, antifreeze, and air temperatures. Resonant transducers are an example of a MEMS sensor that have been attracting a widespread scientific and technological interest in the development of miniaturized chemical and biological sensors. They can be used for the measurement of pressure, acceleration, strain, temperature, vibration, rotation, proximity, acoustic emission, and many others.

In order to characterize this wide range of different devices the laser doppler vibrometry is a valuable technique thanks to his non-contact nature and to the possibility to position the laser beam in every point on the object surface.

With this instrument it is possible to dynamically characterize the systems: to find resonance frequencies, distinguish mode shapes, revealing the existence of all the vibrational modes, nonlinear behaviour and also to investigate the mechanism of mechanical energy dissipation that play a fundamental role in the performance of the devices (quality factor assessment) [3, 4].

2 Measurement System

The system is based on a commercial Laser Doppler vibrometer, in which the optical set-up, the mechanical arrangement and the processing software and hardware were modified and developed to measure vibrations with a resolution in the micro scale.

The main characteristics of the developed system is a very versatile platform, in which laser doppler vibrometry, digital signal processing and image acquisition and processing can work together (figure 1).

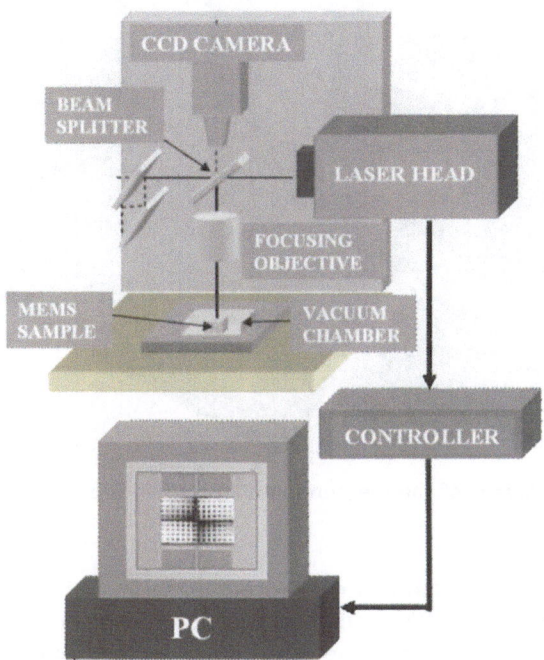

Fig. 1.   Scheme of the measurement system.

In order to scan the sample surface, it is necessary to position the laser beam on the points of a measurement grid built over the sample surface (figure 3), we used two different approaches:

▶ To move the object by means of mechanical micropositioner, maintaining the laser beam in a fixed position.
▶ To move the laser beam across the surface of the MEMS by means of piezoelectric mirrors with the sample in the same position (figure 2).

A fundamental element of our measurement system is represented by the focusing objective that allows us to reach a suitable spatial resolution by reducing the laser spot dimensions. The objective that we implemented is composed by two lenses with a focal length of 30.17 mm corrected for the spherical aberration that reduce the laser spot dimensions allowing a special resolution of 5 μm (laser spot diameter).

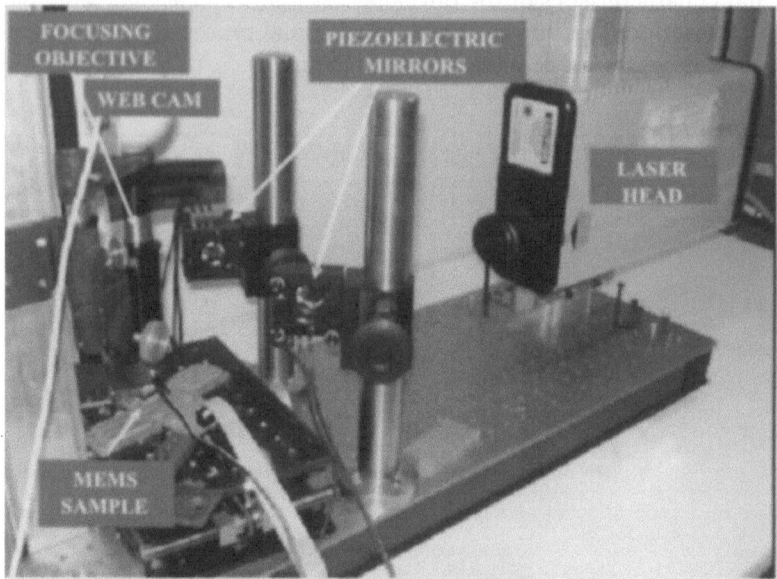

Fig. 2.   Scheme of the measurement system with piezoelectric mirrors to move the laser beam.

Another important element of our system is the post-processing software that was developed in LabVIEW environment. The main characteristics required were the ability to integrate the control of each component (stages, cameras, laser etc) and to make easier and automatic the measurement procedure.

In particular it had to answer to the need of:
  ▶ Managing the images acquisition.
  ▶ Making possible to build the measurement grid.
  ▶ Managing the excitation system.
  ▶ Managing the acquisition parameters.
  ▶ Managing the data post-processing.
  ▶ Managing the measurement procedures:
    ▶ Vibrational analysis.
    ▶ Automatic Q factor calculation.

4   Vibrational Analysis

The measurements were performed on out of plane accelerometers. The sample was excited by using white noise to determine the resonance frequencies and with sinusoidal signal at the identified resonance frequencies.

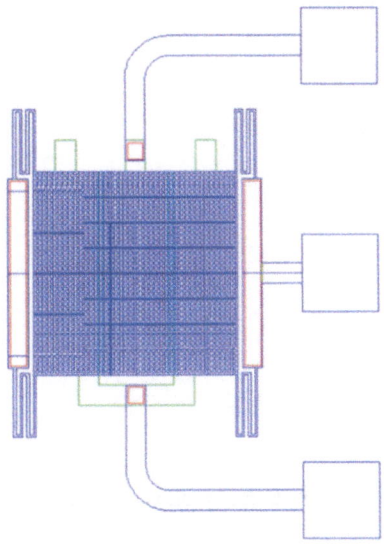

Fig. 3.   Scheme of the out-of-plane accelerometer.

With our system it is possible to obtain the time history, the frequency spectra, and the maps of the vibration velocity and the displacement.

In figure 4 is shown the average spectrum acquired over the measurement points of the grid by exciting the sample with white noise. It is possible to notice the presence of a resonance peak at the frequency of about 35 kHz and hits second harmonic at the frequency of 70 kHz.

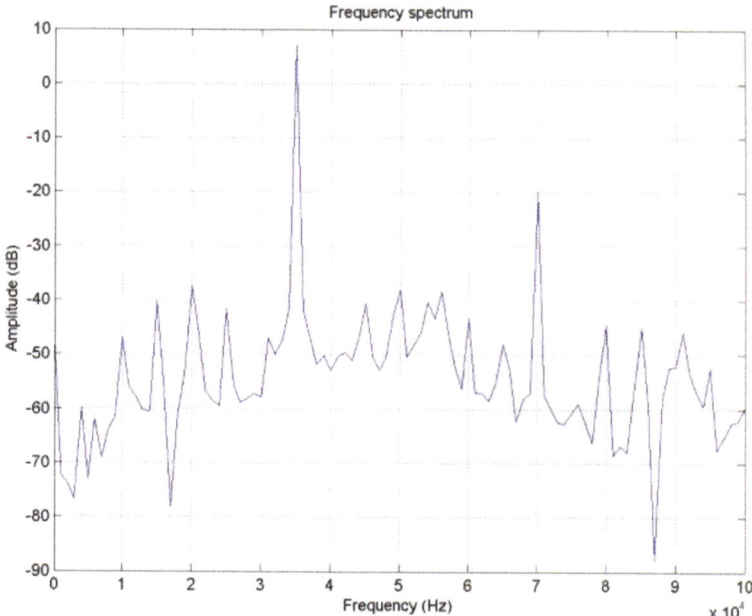

Fig. 4.   Frequency spectrum. Excitation signal: white noise.

In figure 5 is shown the time history obtained by averaging the data acquired over the 25 points of the measurements grid. The sample was excited with a sinusoidal signal at the frequency of 35 kHz. Each measurement point was acquired with a number of averages varying from 4 to 32 depending on the noise level.

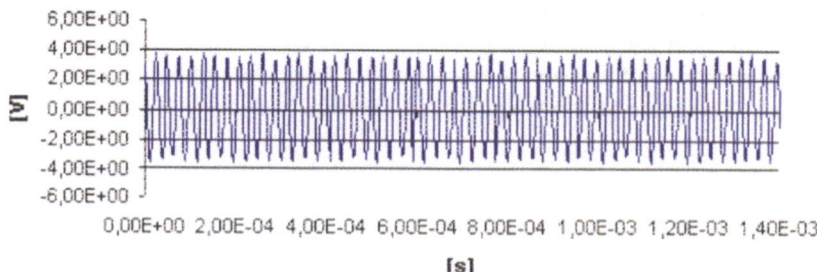

Fig. 5.   Time history. Excitation signal: sinusoidal wave.

In figure 6 is shown the measurement grid that has been built over the sample image by the developed software. The laser beam is positioned over each measurement point by the piezoelectric mirrors driven by the same software.

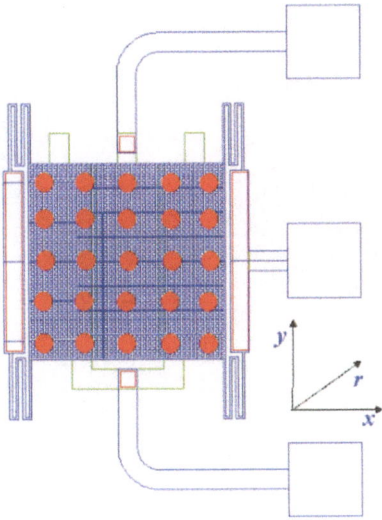

Fig. 6.   Measurement grid.

With the microvibrometer it was possible to measure the vibration velocity and, by post-processing the data to obtain the vibration velocity map and the displacement over the accelerometer surface as shown in figure 7 and 8. It is possible to notice that the higher velocity and displacement are, as expected, in correspondence of the centre of the accelerometer.

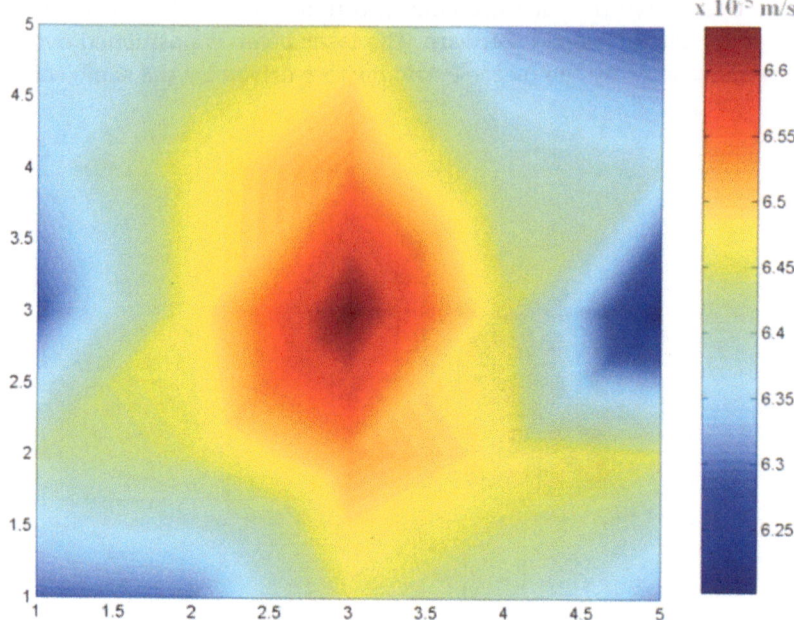

Fig. 7.   Vibration velocity map.

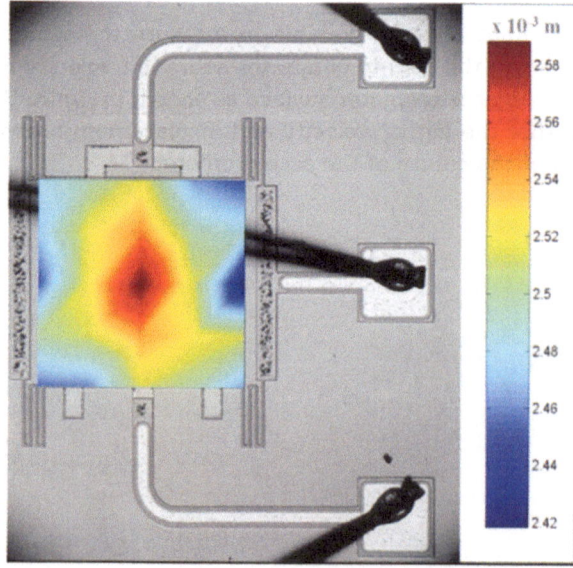

Fig. 8.   Displacement map superimposed on the microaccelerometer.

7 Conclusion

In this paper has been demonstrated the LDV system is a very suitable tool for measurement in the microscale thanks to his non-contact nature and to the possibility to position the laser beam in every point on the object surface.

With this technique it is possible to dynamically characterize the systems: to find resonance frequencies, distinguish mode shapes, revealing the existence of nonlinear behaviour and to calculate the Quality factor of the devices.

In this research work we applied such technology to study the dynamic behaviour of out-of-plane microaccelerometers. We were able to determine the resonance frequency of 35 kHz and to reconstruct the map of the vibrational velocity and of the device displacement. The visualization of the displacement is very useful to verify that the device is moving accordingly to the design specification and detect the presence of anomalies in the MEMS behaviour.

It is also possible to superimpose the obtained map with the picture of the sample acquired by the CCD camera to have an easier visualization of the obtained results.

References

[1] "The MEMS Handbook", M. Gad-el-Hak, CRC press, 2002

[2] P. Castellini, B. Marchetti, E. P. Tomasini "Scanning Laser Doppler Vibrometer for dynamic measurements on small-and microsystems", Proc. of Fifth International Conference on Vibration Measurement by Laser Technique: Advances and πApplications, SPIE vol. 4827, pp.194-200, Ancona, 2002. [4827-10].

[3] "Characterization of silicon micro-oscillators by scanning laser vibrometry", J. F. Vignola, S.F. Morse, B.H. Houston, J.A. Bucaro, M. H. Marcus, D. M. Photiadis, and X. Liu, Journal of Scientific Instruments (December, 2002).

[4] "LDV measurements in MEMS: a sampling of capability", Lewin, Andrew, Proc. Of 2002 PSV users meeting, San Diego, 2002

[5] "Systems with small dissipation", Braginsky, Mitrofanov, Panov, The University of Chicago Press, 1985.

[6] "Guide to the expression of uncertainty in measurement", 1995

Paolo Castellini, Barbara Marchetti, E. P. Tomasini
Dipartimento di Meccanica
Università Politecnica delle Marche
Via Brecce Bianche
60131 Ancona
Italy
b.marchetti@mm.univpm.it

Keywords: MEMS, Laser Doppler vibrometry, resonance frequency

Mobile Vision – Developing and Testing of Visual Sensors for Driver Assistance Systems

R. Preiß, C. Gruner, T. Schilling, H. Winter, Aglaia Gesellschaft für Bildverarbeitung und Kommunikation mbH

Abstract

Visual sensors are going to play an important role in driver assistance systems. The authors present examples for the application of visual sensors. They present an overall architecture for a general technological platform and individually selected aspects. Special emphasis is attached to systematic testing and evaluating. They present a test procedure with virtually generated driving scenes.

1 Visual Sensors in Driver Assistance Systems

1.1 The Role of Visual Sensors in Driver Assistance Systems

Important traffic signs have signalling colours, reflecting coatings of road marking increase safety, and streets are illuminated by spotlights and street-lamps during night. These are only some examples showing that traffic infra-structure is designed and optimised for recognition by the human eye. That is why cost-effective driver assistance systems are necessarily going to be dependent on using such signal sources in an adequate way. In addition, the risk of mutual influence is reduced in mass application by waiving active com-ponents (e.g. radar), and another cost benefit is achieved. Thus, visual sensors are going to play an important role in future driver assistance systems.

Intensive research in the field of image processing opens perspectives for a new handling of vehicles by man. The vehicle acquires with its sensor technology knowledge about the environment, reprocesses it, links it intelligently, and presents it to the driver in an appropriate way. Due to its great number of pos-sible applications and technical variants of realisation, it is a great challenge to make an essence out of this mass of theoretical ground work and to develop a product ready for the market out of this. For this, you have to make the step from theory to practice and to manage the tremendous difference between pro-totypes and mass production. What does this cost? How stable is it? There are many questions emerging on the way, which have to be answered.

1.2 Application Examples of Visual Sensors

We would like to show the application of visual sensors in driver assistance systems with the help of four examples.

At lane departure warning (see figure 1), images of a camera system turned to the front are analysed in order to generate a model from the lane ahead by visual analysis of the road marking. Out of this, the system generates estimation about the lane intended to drive within the next seconds and minutes. Simultaneously, the actual driven lane is always carried along by analysing the vehicle's CAN data. A warning signal can be produced by continuously comparing such data in order to draw the driver's attention to an unintentional deviation from the lane.

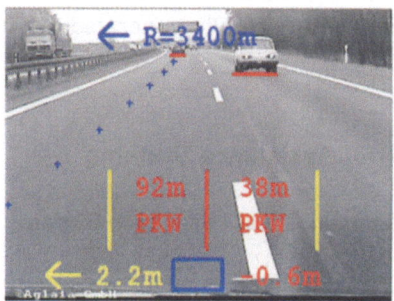

Fig. 1.   Lane departure warning.

The included lane identification functionality is a requirement for many other applications, like e.g. the automatic triggering of a braking process when obstacles appear on the lane, the alignment of the front lights to the course of the lane, or a semi-automatic support when driving bends.

Fig. 2.   Traffic sign recognition.

Traffic sign recognition (see figure 2) has a sensor arrangement very much related to the above mentioned one. A camera turned to the front is used here too, in order to recognize traffic signs located near the lane. The recognized traffic signs are saved. In this way, the current applicable speed limit can be signalised to the driver by analysing the speed limit signs. The recognition of stop signs or give way signs can help compensating the driver's inattention and observing the give way rules applicable at the next junction.

Unlike the above said, blind spot detection (see figure 3) requires a camera for this, which can be placed into the side mirror, and which is focused on the blind angle. The system permanently analyses the motion dynamics in this area and fades in a warning light in the rear mirror as long as there is an object in the blind angle. In addition, the calculated motion dynamics can help to warn of overtaking vehicles approaching at very high speed.

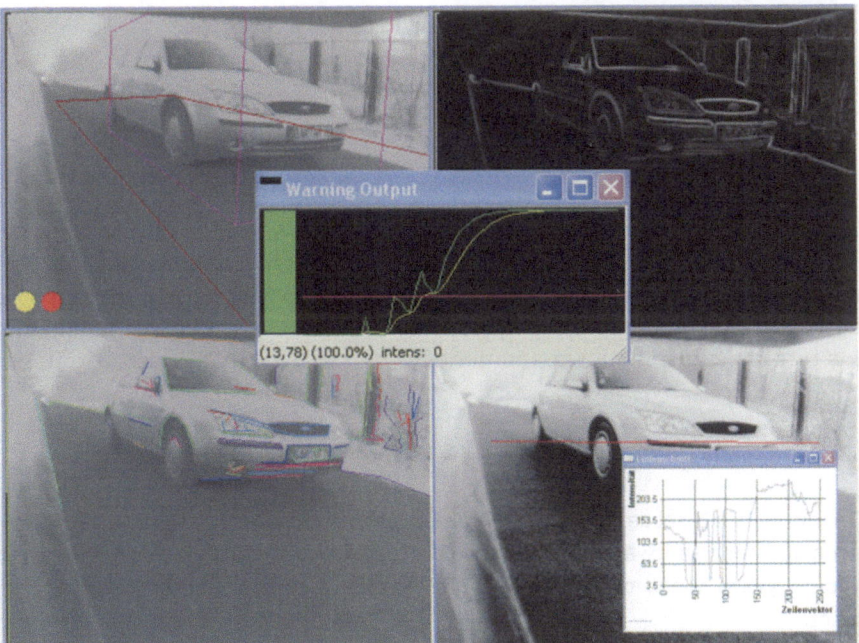

Fig. 3.   Blind spot detection.

The principle applied for this (expanding a man's field of vision by a camera) is used for the backing camera too (see figure 4). A wide-angle camera mounted at the rear is focused on the rear bumper's area.

It releases the driver from the agony increased by more and more rounder and higher rear contours to back either like flying blind using the bumper's crushability or wasting space for shunting.

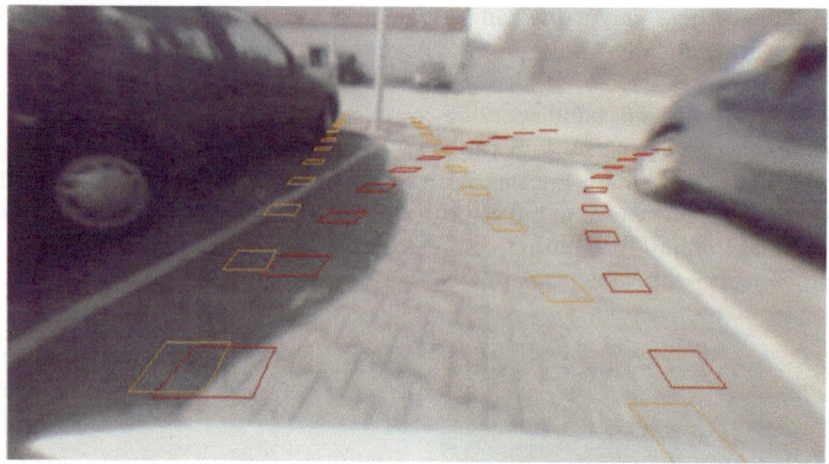

Fig. 4.   Rear view system.

During this process, the wide-angle lens' distortion is corrected by the software automatically. Subsidiary lines faded in show the driver valuable distance information and support the parking process. Depending on the current steering wheel's turning angle the tyres' track, which would occur if maintaining this turning angle, is shown additionally.

1.3   Questions Raised during the Development of Visual Sensors

The entry of visual sensors in the vehicle is connected with some questions, to some of which there are currently no definite answers. It will be decided in the coming years, what answers and which technology will succeed in mass application.

If you just look at one single of the above applications at first, you will find already a profusion of technical variants when choosing an appropriate camera system. A binocular system with two lenses is able to extract better depth information from the scene compared with a monocular system and thus, it facilitates three-dimensional modelling for the following algorithms. This advantage is faced with higher costs and a more complicated assembly. Apart from the optical features (like e.g. the aperture angle), it is also important when selecting the camera whether a colour camera is used or grey scale values are sufficient. Requirements to the sensor's night view capability as well as general prescriptions from the automotive sector are increasing the selection conditions further.

If several applications of the above kind shall interact, it will automatically raise the question of common utilization of (here in particular the cameras') resources. With how many cameras will a future car be equipped? How many cameras will be required for an all-round view? At what places will they be installed? How does this go together with the designer's ideas? What is the enduser's acceptance regarding various realisation variants?

These open questions are already reason to believe that the integration of visual sensors will have great dynamics in the future and that short innovation cycles are to be expected. These considerations have a direct impact on the development of visual sensors. For using components evolving from this in different application scenarios, they should be highly reusable. In addition, the principles of rapid prototyping will be useful in order to realize various alternatives within short time and to develop those very promising further. Development environments, which permit a systematic evaluation, will be particularly important for this.

2   General Framework for Embedded Visual Algorithms

The positioning of suppliers of visual sensors within the automotive industry's gear is in the state of flux. Suppliers are faced with the question how intensive they should concern themselves with this new technology. You cannot assign visual sensors unambiguously to just one automobile component. You may combine them with headlights and integrate them into bumpers and rear mirrors. In the interest of reliable and compatible systems, it is wise to generate technical platforms for visual sensors, which can be used in many ways, and which can be integrated into different environments.

Moreover, such a technological platform makes a common utilization of resources possible. You may use the same camera for various tasks, and storage and computing performance can be concentrated on applications running in the foreground.

Visual sensors are a new issue for many suppliers. We would welcome if it could be possible to avoid proprietary isolated applications from the very beginning and to develop common interfaces similar to the aims of the AUTOSAR initiative.

We outline the architecture of development platform visual sensors below and present reusable components.

2.1 Architecture of a System for Rapid Prototyping

A modular software design is natural. A standard regarding the exchanged data should exist for the separate modules to be able to communicate with each other. This concerns both the semantic level (which terms and objects are distinguished) and the concrete representation on the implementation level (data format). The principle way of exchanging data is part of the design decisions to be made early. In this case, we suggest the data flow orientated approach, at which three types of components exist:

▶ **Data sources** produce data and send them along the determined variable data flow.

▶ **Data processing stations** receive data from different sources, modify and forward them.

▶ **Data sinks** receive data and display them usually to the user or save the data.

We recommend a bi-directional approach in general, at which the data flow can be actuated from data sources when sensor data are present, as well as into the opposite direction by requesting data through data sinks, if an output device has to be operated at a constant data rate for example.

A system's overall architecture for rapid prototyping with visual sensors can look like the one below:

You can differentiate between four essential components. In the sensor layer, the user may select from existing sensors by Plug&Play. These sensors can be switched on/off individually, and they produce a data stream of independent sensor data at first. This data stream is then bundled by time stamps in a synchronized way and is made available to the second layer. This layer too is put together by Plug&Play from a database of components compiled beforehand. Additional connections between these processing modules may be determined here individually. The separate connections are associated with data types, like e.g. CAN data, image data, or intermediate data of calculations.

The third layer serves to output the results, usually displayed on a monitor. Alternatively, generated data may be fed back into the CAN bus. You may configure this layer too as you like it by Plug&Play.

A fourth component is working in the background. You can record with it both sensor data and the results, in order to reproduce the same situation later and if necessary compare the results of new algorithms with those of older configurations.

As we have already mentioned before, you can already see at this schematic level that you have to determine the data types for communication between the separate components.

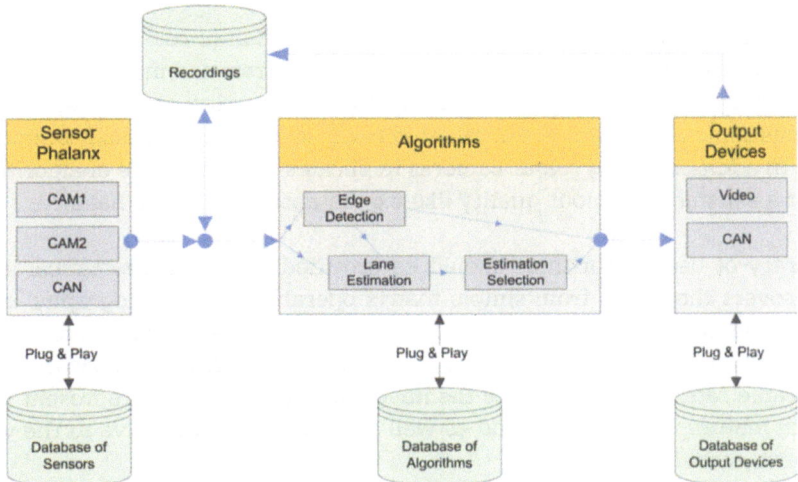

Fig. 5.   Software architecture for rapid prototyping of visual sensor applications.

2.2 Examples of Reusable Software Components

Let us take individual components as an example, which are playing a role in the development of visual sensors.

Currently available cameras usually have a maximum bit depth of 12 bits, which permit 4.096 different brightness values. Using logarithmic characteristic lines permits a great scope of dynamics, but at the expense of the resolution between the discrete brightness values. That is why; it is necessary to use the characteristic line optional. In most cases, you have to adjust the camera additionally, i.e. to adjust the camera's set-up parameters (like e.g. the exposure time) to the respective situation. Since this adjustment can be done only based on the camera images present beforehand, which are dependent on the set-up parameters' history, this problem is a question of a control circuit. You can think of various solutions, which also have their relevant justification, for the requirements on the camera's control vary depending on the application. If the image is shown to the user directly there should be no brightness oscillations, but the system should react as fast as possible to the change in brightness. However, if on the other hand images are becoming part of the image processing algorithms, it might be necessary to adjust as slowly and smooth as possible, so that extreme jumps in brightness do not occur between the single images. In this case, the algorithms could get problems in the follow-up of recognized features.

Another algorithm is the Bayer pattern reconstruction in colour cameras with Bayer Pattern, at which RGB values are calculated from the mosaic like spread colour information for each pixel. Especially when implemented in embedded systems some algorithms are eliminated because of their increased computing time. In addition, using colour cameras implicates another range of algorithms to get a satisfactory colour quality like e.g. an automatic white balance.

A library of basic algorithms should be provided on the mathematical level. This covers the range from simple matrix operations, via homography, up to the implementation of complex processes, if problems arise with auto calibration for example. The aim is to carry out a camera self-calibration based on just one frame rate, i.e. to determine the image parameters, which represent their optical features on the one hand as well as their position relative to the vehicle on the other hand.

Statistical methods are used too. A hundred percent safe procedures are seldom in image processing. Image recognition is often connected with a certain probability. However, in order to reach reliable results you can work with two supporting methods. On the one hand, the multiparadigm approach suggests itself, at which at least two different processes are working on the same input data and are synchronizing their results afterwards. On the other hand, hypotheses are often used, i.e. the algorithms establish hypotheses at first, to which probabilities are already being allocated. Lists of hypotheses are generated by combining evaluations of single images, which can be analysed statistically in order to extract the most probable hypotheses.

All these problems can be isolated and looked at as a black box with specified in- and outputs. If necessary, several solutions can then be provided for each problem.

2.3   Software Development Structure

The research process in the field of visual sensors is rapidly moving forward. New ideas are swirling at its front tamed and bundled in the maelstrom of research. Stabilized procedures establish themselves in the rear area and become the status quo. Software has a similar structure. New developments are changing quickly, are revised, and some of them are taken over into the core. The core code has stabilized by having reused it, and a proven structure has formed in the course of time.

Classes like e.g. image format, image data, CAN message, mathematical methods, models (matrices, homographies, transformations, ...), and much more are part of this structure. You cannot measure the value of such classes in code lines, because exactly that selection process is essential, which took place before and where alternatives have been tested and turned down. It is important to be aware of the fact that the code is only a small part of that knowledge, and that the essential part is the ideas, experiences, and concepts behind it.

You can design the core in such a way that it may be used both for prototypes (however, you need an additional layer for the user interface for this) and for its implementation in embedded systems. It is recommended working with prototypes and using their short development cycles.

A requirement engineering tool, like e.g. DOORS, IRqA, CaliberRM, or Xtie is wise for taking down the requirements and target definitions. A performance specification is better structured by systematic collection, and some questions are answered before the implementation cycle starts. The time you invest here pays to sped up implementation.

2.4   Common Hardware Components

When selecting the hardware architecture for visual sensors it is of crucial importance that you have to reach a very high data throughput and that some algorithms require an additional buffering of single images. That is why the internal storage has to be dimensioned sufficiently and the access to external storages has to be fast enough. The camera, the CAN bus, the video output, and if necessary individual signals have to be supported for user interaction, because they are the essential in- and output interfaces.

Manufacturers (like e.g. National Semiconductor Inc.) offer promising solutions in the field of CMOS sensors, which have already built in algorithms to improve image quality or for image compression. When integrating a camera, we suggest the Firewire interface for prototypes or distributed systems, which is suitable for automotive according to the IDB-1394 standard, and which in particular makes a synchronous triggering of several cameras over one single connection possible.

Fig. 6.   Aglaia camera components.

When selecting a camera, we suggest using CMOSs cameras for the following reasons among others:

▶ CMOS technology is becoming a mass product increasingly with a corresponding impact on pricing.

▶ CMOS sensors have high dynamics and permit the definition of image sections as a region of interest.

▶ CMOS technology is easy to handle and facilitates circuit integration.

It is advantageous for the integration of CAN data to include an intermediate level converting manufacturer specific CAN messages into generally usable symbolic designators.

3   Tests and Evaluation

Unlike research orientated prototype development, you should allow reliability and robustness a central role in order to market visual sensors successfully. Here, a special feature of image processing algorithms comes into it for they often have a heuristic approach in their core, so that you can establish the results' quality first after having carried out numerous series of tests even at an absolutely perfect implementation.

The reliability of recognizing the lane is depending on the weather and on the light conditions for example. Reflections on the lane can make recognition more difficult. Road marking may be interrupted at junctions or building site areas. You can face this challenge in different ways. We would like to present two components here.

3.1 Quality Assurance through Environment Models with OpenGL

Using records of everyday driving scenes for the evaluation of systems with visual sensors, the problem of an unambiguous collection of the target results arises. If traffic signs shall be recognized for example, the actual traffic signs appearing on the records have to be collected in an appropriate way at first. According to the current state of the art, you can do this in a reliable way manually only, by which an evaluation becomes very cost-intensive. Using other data sources (e.g. laser aided object recognition or from maps with GPS data) is limited too.

You avoid those disadvantages with the following simulation test method. For this, an artificial 3D-model is generated at first. In this process, you can select various elements from a library, like vehicles and their movements, lanes, traffic signs, trees, etc. for example. Their positions and shapes are parameterised. Those parameters (e.g. the position and type of the respective traffic signs) do directly correspond to the target values. This 3D-model is represented by OpenGL and is presented to the image-processing algorithm in such a way, as if those data are coming from the sensor directly. You can reach a realistic representation through further effects like textures, shadow casting, mist, and much more (see figure 7 and figure 8).

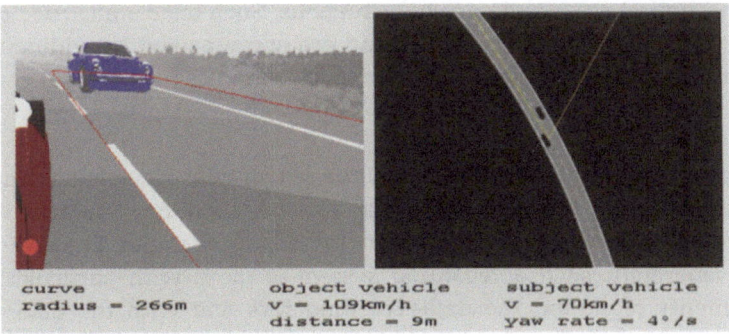

Fig. 7.   OpenGL simulation for blind spot application.

By this approach, the values established by the algorithm can be compared directly with the target values and be evaluated statistically reliable. This method is completely reproducible, so that even different algorithm can compete against each other.

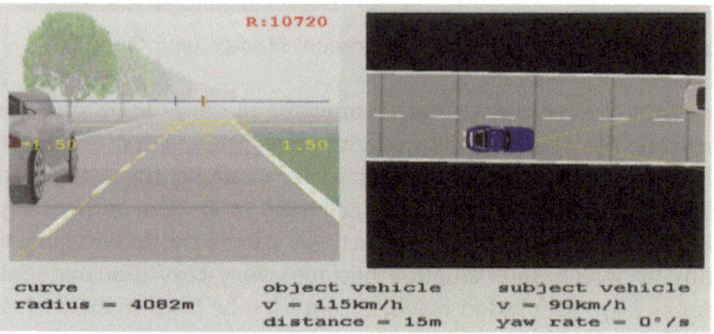

Fig. 8.   OpenGL simulation for lane detection.

A particular advantage of this method is that it can be completely atomised, and that you can quickly build up a database as large as one likes by modification of the parameter sets.

3.3   Record and Playback of Driving Scenes

The above-described method allows a very systematic approach. However, it includes the disadvantage that the examined (abstract generated) driving situations are unable to reproduce reality completely. That is why they are supplemented by tests, in which the initial data are records of realistic scenes including typical cases and extreme situations. Such tests are carried out during the whole development process and are accompanied by additional field tests, in which a great number of test vehicles are used for adjusting and validating the results.

In this process, it is important to record the sensor data occurring during a test drive in such a way that a later repetition in the laboratory is as much as possible equivalent to the original situation. In addition to the task recording each sensor separately, such a recording system is faced with particularly high requirements regarding the data-recording rate and with the data volume resulting from this. You should particularly focus on the sensor data's synchronous playback for the sensor signals' temporal relation to each other has a strong influence on the algorithms.

Driving scenes with data from the cameras, the vehicles (CAN), and the current calibration etc. are collected via such a recording module and transferred to a server. Such data collection is a requirement for quality checks and evaluations carried out continuously.

3.3 Conclusion

The data flow orientated development environment Cassandra® of Aglaia GmbH for Rapid Prototyping with a stabilized core of image processing algorithms and data structures and the MobileScope module for synchronous recording and playing back of sensor data is a good basis for a economical development of systems with visual sensors and their product scaling to embedded systems. An essential foundation stone for automatic evaluation is laid by the test component for artificial creation of driving scenes. Further hardware components for a general technical platform for visual systems are currently prepared and may be used in various fields of application.

References

[1]   AUTOSAR, Automotive Open System Architecture, http://www.autosar.org/

[2]   HIS, Herstellerinitiative Software, http://www.automotive-his.deReference

Roland Preiß, Christian Gruner, Thomas Schilling, Harald Winter
Aglaia Gesellschaft für Bildverarbeitung und Kommunikation mbH
Tiniusstr. 12-15
13089 Berlin
Germany
mail@aglaia-gmbh.de

Keywords: Driver Assistance Systems, Visual sensors, Hardware framework, CMOS cameras, OpenGL Test Scenario

Silicon Capacitive Absolute Pressure Sensor Elements for Battery-less and Low Power Tire Pressure Monitoring

J. Thurau, J. Ruohio, VTI Technologies Oy

Abstract

The sensor industry has seen a drastical change due to replacement of mechanical structures with MEMS technology making sensors smaller and more cost effective [1, 2]. With higher integration, shrinking size and automated processes the quality and robustness was enhanced dramatically in the same time to meet high volume automotive applications demands. This process can be seen as an ideal synergy due to the high volume production that leverages the high capital and fixed costs. The lessons learned from this roadmap was brought back to a simple looking device – a capacitive absolute pressure sensing element as bare die. This article will describe the way how this element was shrunk and optimised for the next generation of low power Tire Pressure Monitoring and Systems without battery.

1 Evolution of Capacitive Pressure Sensor Elements Due to TPMS Requirements

A group of engineers started in Finland in 1980 a development of pressure sensors targeted for weather measurement application. Today their work is found in meteorological radiosondes manufactured in high volume and in high accuracy barometric instruments. After 20 years of using this sensor technology in highly sophisticated high accuracy systems the same technology was applied in a low cost, mass producible, robust, low power pressure sensor for an extremely harsh environment in automotive tires.

In a cutting edge product like battery-less TPMS critical design parameters are involved. In the application the following factors were identified:

- ▶ **Size:** The entire system needs to be extremely small since it should be mounted in the tire in several applications. Tire tags have been presented with the size of a 1-cent coin.
- ▶ **Costs:** The added value perceived by such systems by the end user is relatively small. Since legal regulations in the U.S. will most probably

lead to a 100% equipment rate and 4 to 5 measurement units are required for one car the price per unit is constantly under surveillance.

▶ **System mounting and integration:** For the less weight (and therefore small size) demand it is necessary to optimize each component. The weight of the TPMS unit in the wheel needs to be compensated since it is positioned unsymmetrical in the wheel leading to dynamic disturbances.

▶ **Robustness against overpressure and temperature stability:** One option is that the sensor in the system will be vulcanized into the tire – proof pressure requirements of more then 32 bar are seen as requirement as well as exposure to a very high temperature, all this without a change in calibration or reliability.

▶ **Centrifugal acceleration:** The measurement shouldn't be effected by centrifugal acceleration or shocks from the road up to 1000 g.

▶ **Media compatibility:** The tire is sealed so that there isn't typically any exchange of air. Outgasing of several more or less harmful agents could lead to severe damage of electronic – one is sulfur.

▶ **Stability over lifetime:** Over the specified lifetime of 10 or 15 year a deterioration of the sensor functionality isn't acceptable.

▶ **Low power consumption:** In battery application the charge is limited and should guarantee a lifetime of more than 10 years. Battery-less systems are even more restrictive since the entire power needs to be transmitted by induction to the tire module and form factors of antenna and their positioning are pretty much dependent on the necessary energy that needs to be transferred so that this variant is defined as ultra low power application.

▶ **Signal ratio:** Due to the low power consumption requirement signal conditioning becomes critical. A comparatively large sensor element signal has the advantage to be much better signal conditioning including less vulnerability against EMC.

▶ **Variation and temperature behavior:** Small part to part variation and excellent over temperature behavior reduces the effort that needs to be taken into energy budget for signal conditioning.

2   Concept to Manufacturing: Small Improvements in All Details

Starting with the most important parameter for the entire sensor element design several simulation were made to define the best size which was still fulfilling the requirements of all other functions. It turned out that a footprint of 1,4 x 1,4 mm² brought the best optimisation. Smaller size would have had the disadvantage in the system manufacturing that the part couldn't be handled anymore. The height was determined to be 1.0 mm as a good compromise. Thinner devices wouldn't have had the required stiffness to compensate mechanical stress that would have effected the measurement accuracy. Thicker elements would have meant bigger dicing grooves which would have reduced the number of elements per wafer.

Fig. 1.   Image of VTIs APS4.

In the simulations determining the size the most important factor was to keep a high dynamic measurement range. Hence the small size this range was kept as high as 7 pF for 0 bar to 14 pF for full scale pressure. This was achieved by an extremely small distance between the two capacitor plates of nominal 1 µm. The passive parallel capacitance was kept small by using a thick glass layer between the thick and thin silicon layer with silicon feedthrough; this provides a good active to passive capacitance ratio. The transfer function follows the equation:

$$C = C_{00} + \frac{C_0}{1 - \frac{p}{p_0}} \qquad (1)$$

whereas C_0 is large compared to C_{00}. The realised values are in the range of: $C_{00} = 1$ pF and $C_0 = 5$ pF. Figure 2 is showing measurement results of first prototypes, an 8 bar element.

Reflecting the over temperature uniformity of the transfer curve provides the perspective that a temperature compensation wouldn't be necessary for this type of element. The major two reasons for this positive behaviour are the above mentioned small C_{00} impact as well as the fact that silicon is used as a conductor only so that the temperature dependency of other semiconductor properties isn't effecting the over temperature behaviour. This advantage needs to be preserved by using appropriate signal conditioning and mounting technologies which should avoid thermal related stress to the sensor element.

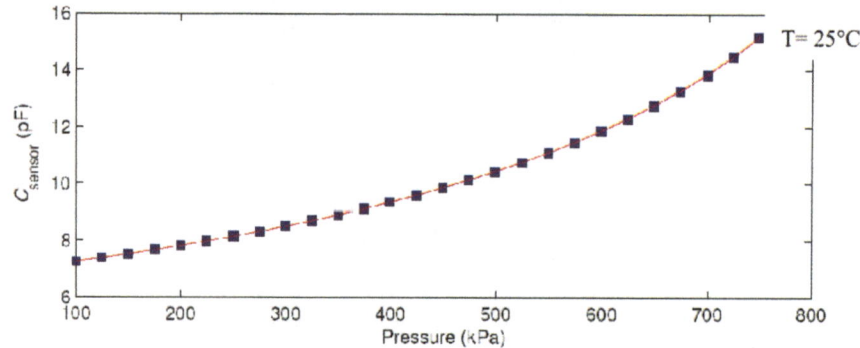

Fig. 2.   Transfer function of APS4 of first prototypes (measurement results).

Tackling the long-term stability the anodic glass bonding needs to seal for life-time. Therefore good process controls in production are required. VTI has gained experience in glass bonding technology for close to 20 years in series production. Glass is therefore a well known material but has also some specific characteristics like outgasing and absorption of residual gas on the glass surface which would lead to change of the internal pressure. This effect was addressed by adding wet etched gas pockets connected to the reference pressure volume enable good long-term stability. The difficulty in the given design was to find any space in the optimised element structure. The positioning of that vacuum reservoir is shown in figure 3 schematically and in figure 4 as a cross section in silicon.

A major application driven design rule was that the connecting pads should be on the lower side of the element so that the part can be directly mounted to a carrier like a PCB or ceramic. Another advantage is that the size of the element is reduced by the pad size that was consuming space on the surface and was difficult to protect against aggressive media. Figure 5 shows older pressure sensor designs, APS2 and APS3, which had bonding pads on the upper side.

The design of APS4 with two different terminals on the lower side was realised by glass isolator in the middle of the element. The glass is also structured in a way that the offset capacity value is kept as small as possible.

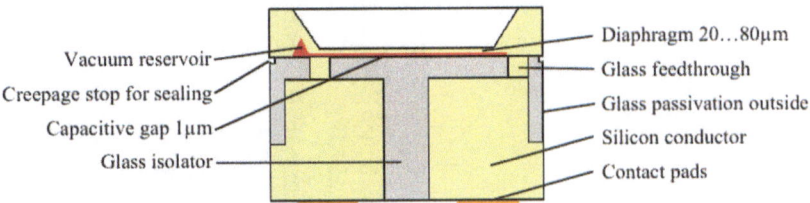

Fig. 3.   Schematic cross section APS4.

The diaphragm thickness is determined by the full-scale pressure range. The design is dimensioned in a way that sensor elements from 1 bar (20 μm) to 30 bar (80 μm) would be realised. By using wet etching the height of the diaphragm can be processed quite accurate. The gap is always kept at 1 μm to remain with the same dynamic range for all the potential variants of this concept. The diaphragm thickness is determined by the full-scale pressure range. The design is dimensioned in a way that sensor elements from 1 bar (20 μm) to 30 bar (80 μm) would be realised. By using wet etching the height of the diaphragm can be processed quite accurate. The gap is always kept at 1 μm to remain with the same dynamic range for all the potential variants of this concept.

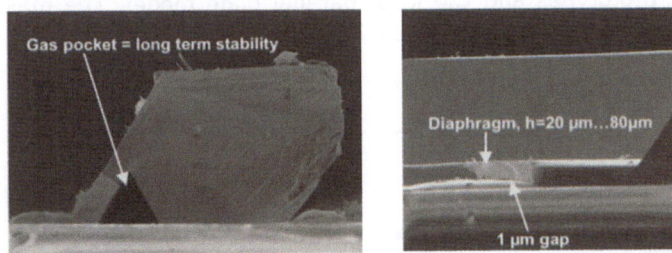

Fig. 4.   Vacuum reservoir for long term stability (left); diaphragm thickness and distance between capacitor plates (right).

Underlying the design of the diaphragm was the requirement to avoid any signal changes caused by centrifugal force as well as mechanical shock caused by rough road surfaces. For that reason the boss shown in figure 5a was removed scarifying some linearity of the transfer function.

Further shape optimisation of the diaphragm border area brought a good compromise for the transfer function.

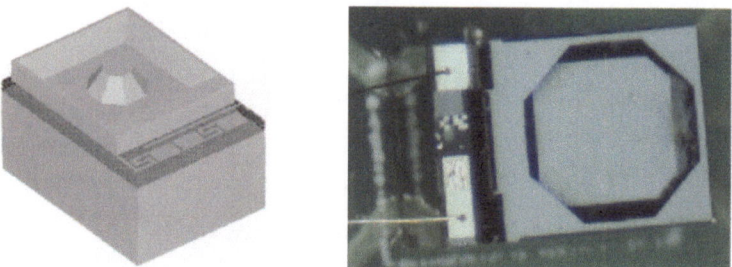

Fig. 5.   APS2 with boss (left); APS3 with bonded contact areas (right).

3   Signal Conditioning

The sensor behaviour is very good in line with the parallel plate model according to the formula:

$$p = p_0 \cdot (1 - \frac{C_0}{C - C_{00}})$$ (2)

Figure 6 shows that there is only a small deviation between the transfer curve of first sensor prototypes and the ideal parallel plate model. The maximum deviation can be read as approximately 7 mbar.

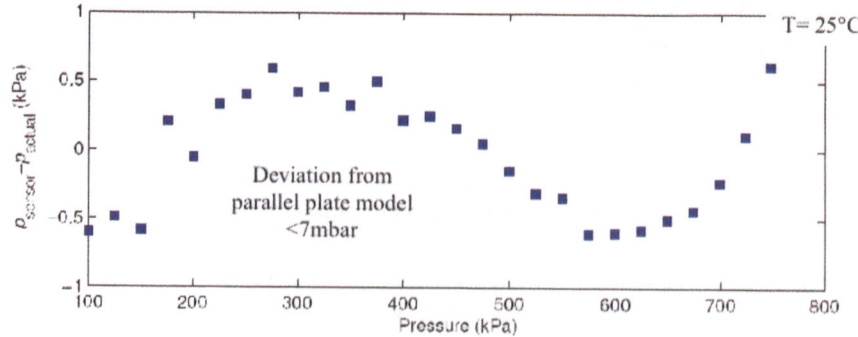

Fig. 6.   Small difference between parallel plate model and sensor output.

With capacitive and piezoresistive measurement principles there are competing concepts on the market. Whereas the piezoresistive principle is in today's TPMS systems in series production it didn't meet the requirement of battery-less systems anymore due to the fact that the power consumption is much higher and for a realisation sleep modes for the measurement acquisition is mandatory. The power consumption for the APS4 sensor element is quite low since the parallel resistance to the capacity higher than 10 GΩ leading to very low leakage currents.

The signal conditioning can be done by several concepts whereas the typical switched capacitor technology is extremely accurate and robust against disturbances but consumes too much power to be an appropriate approach for this ultra low power application. The good over temperature stability of APS4 pressure sensor element helps to avoid a reference capacitor in element technology. The sensor capacity can be connected directly to the oscillator circuitry of the ASIC interface.

4   Implementing APS4 into Application Structures

The vertical groove glass process enables a MEMS component to be a reflow solderable SMD component. This leads to easy positioning of the sensor element with other SMD components so that wire bonding isn't necessary as interconnection technology at all (figure 7).

The media compatibility can be realised by using an appropriate sealing material that covers the element up to the glass passivation so that only the glass passivation and the silicon diaphragm are exposed directly to the harsh environment. This gives an advantage compared to other structures which are exposed on the surface of the element.

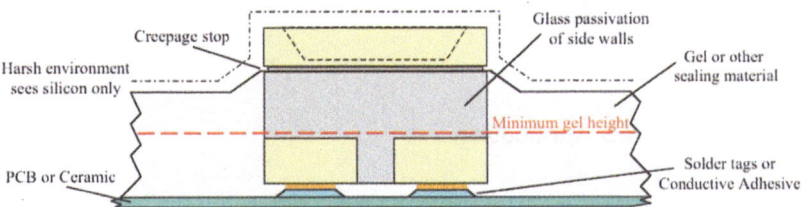

Fig. 7.   Proposal for mounting of APS4 on PCB or ceramic.

The sidewalls of the sensor below the glass passivation (electrodes shown in yellow) have to be protected since foreign media with large dielectric constant could change the parasitic capacitance of the element on the bottom edge of the sensor element sidewalls – the minimum gel height is shown as a red line. While using a potting material for this protection a potential side effect could be that the material has a tendency to creep on top of the element due to the surface tension. This would imply more mass on the diaphragm which would make the sensor more vulnerable for centrifugal forces or shocks like the boss design. This can be avoided by the creepage stop, a small groove on all four sides of the sensor. The rest of the electronic can be positioned under the protective potting also.

Without having potting material on the surface of the diaphragm the acceleration sensitivity is in a reasonable level that can be neglected according to:

$$p_{eff} \approx P_{Si} \cdot h_{dia} \cdot a \qquad\qquad (3)$$

where as a is the acceleration, h_{dia} is the diaphragm thickness [20...80 µm], and P_{Si} is the density of silicon.

The design was very much optimised for small size and high dynamic range so that there is one parameter that needs to be contemplated while mounting the device to the application structure as well as for the treatment in the logistic chain. Due to the small capacitor plate distance of 1 µm the maximum acceptable ESD voltage level is less or equal than 100 V. This fact makes special requirements in the mounting area necessary like for MOSFET structures. After mounting and encapsulation this special precaution isn't necessary anymore in application.

The sensor element was tested in a set-up where vulcanisation conditioned were applied to the sensor element. The conditions were 185 °C for 1 h under 32 bar pressure. The functionality and calibration of the sensor remained unchanged after this test.

5 Trends in TPMS Application

The benefit of having the correct tire pressure regarding optimal braking distance, less tire wear out and unexpected tire break down was not only shown in several papers [3, 4] but also proven by a series of accidents driving this topic to public awareness. Therefore the market is demanding the introduction of TPMS for the following application classes: Passenger cars need a full scale pressure range of 5 bar; light trucks are defined for 8 bar whereas truck would need sensors with full scale pressure values of 12 bar to 13 bar. Elements for all three pressure ranges will be available from VTI.

Indirect pressure measurement as shown in [5] was one of the attempts to cover unexpected pressure loss in one of the wheels hence it seems to have the disadvantage of being unable to identify equal conditions of too low pressure in all four wheels. Therefore it is most likely that direct pressure measurement in all four wheels will become a standard. Based on the minimum requirement for measuring the actual pressure in all wheels several additional information is driving the development of more sophisticated systems. The tire position identification is definitively a basic demand also since otherwise different pressures for front and rear tires can not be handled. The information can be used beneath under-pressure warnings also for the adaptation of other systems like suspension and braking.

One trend shows that the tire makers will gain more and more responsibility for their products. The most significant option is the traceability for legal purpose which would require a fixed pair of tire and intelligent sensor that could be realised by molding of the pressure measurement unit into the tire. This leads to a battery-less system since molding of batteries into tires would be difficult to realise. A package with low and high frequency data transmission integrated with MEMS sensor provides the working solution without battery. Silicon based pressure sensor elements with capacitive principle are key components when these type of systems are designed due to their lowest power consumption.

5   Other Applications

Robust sensor elements are required by other application also where media compatibility is significant. VTI has the capability to expand the shown concept to derivate products with a full scale pressure range between 1 bar and 30 bar. There are several application specific possibilities to package a simple sensor structure as it was shown in this paper. The signal conditioning and communication depends in the same way as the packaging on the application.

References

[1]   Heikki Kuisma, 'Inertial Sensors for Automotive Applications', Transducers '01, Munich, Germany 2001

[2]   Heikki Kuisma, Tapani Ryhänen, Juha Lahdenperä, Eero Punkka, Sami Ruotsalainen, Teuvo Sillanpää, Heikki Seppä,'A Bulk macromachined Silicon Angular Rate Sensor', Transducers '97, Chicago 1997

[3]   MacIsaac Jr., J.D., and Garrott, W.R., "Stopping Distance vs. Tire Inflation Pressure", DOT Docket, Document NHTSA-2000-8572-70.

[4]   3. Collier, B.L., and Warchol, J.T., "The Effect of Inflation Pressure on Bias, Bias-Belted and Radial Tire Performance", SAE Paper # 800087, 1980.

[5]   F. Gustafsson, M. Drevo, U. Forssell, M. Lofgren, N. Persson, and H. Quicklund, "Virtual sensor of tire pressure and road friction", SAE paper 2001-01-0796, 2001

Jens Thurau, Jaakko Ruohio
VTI Technologies Oy
P.O. Box 27
01621 Vantaa
Finland
Jaakko.Ruohio@VTI.FI, Jens.Thurau@VTI.FI

Keywords: capacitive absolute pressure sensor, battery less tire pressure measurement systems, bulk micromachining, low power consumption, size optimization for system integration

Automotive Applications for the ALASCA Laser Scanner

B. Nitsche, Hella KG Hueck & Co.
R. Schulz, IBEO AS GmbH

Abstract

Driver assistance systems provide road users with enhanced convenience and safety. Sensors for detecting and assessing traffic situations are readily available to the automotive end-customer. These sensors, however, support first generation applications. In this paper an advanced sensor technology together with its applications is presented. The Automotive LAser SCAnner ALASCA® supports applications such as Automatic Emergency Brake (AEB), Stop and Go, Precrash, Parking Assistance, Pedestrian Detection and close range monitoring for commercial vehicles (e.g. "Turning Assist"). These applications are based on the high efficiency of detection and tracking of the objects in the proximity of the car before a collision occurs.

Furthermore, ALASCA® has the potential to support the ambitious plans of the car manufacturers regarding automatic braking and steering of vehicles and accident-free driving in road traffic.

1 Introduction

As a result of the evaluation of road accident statistics, special interest is focused on distance RADAR/LIDAR alert and emergency brake systems. Advanced driver assistance systems (ADAS) have the potential to reduce the number of accidents by monitoring the driver and the vehicle as well as the vehicle surroundings. The need for the development of ADAS functions is backed by the 9/2002 EU initiative "eSafety" which includes an enhanced action plan to enforce the implementation of new technologies for active safety systems and advanced driver-assistance systems.

For the reduction of the number of fatalities and injuries in public traffic, it is necessary to introduce active safety applications to assist the driver in dangerous driving situations. The long-term objective of full collision avoidance is still a vision at the horizon of future developments. Some ADAS products available in the near future are governed by the philosophy to assist the driver in situations when the crash is imminent.

These collision mitigation applications will help to reduce the consequences of crashes to all collision partners. An important prerequisite for such systems is a sensor for object detection and tracking, which has a high resolution for the localization of the colliding obstacle, a precise tracking capability and a high degree of reliability under all driving conditions.

Hella KG and IBEO AS formed an industrial partnership with the intention to develop the ALASCA® laser scanner, which serves the requirements of automotive applications. In the following the applications served by the ALASCA® will be described in detail. Special attention will be paid to the ACC Stop & Go and Automatic Emergency Brake (AEB) applications.

2   Technical Principals of ALASCA

Prior to starting the operational development of the ALASCA®, the project partners thoroughly discussed the interdependencies between the automotive applications, their functional requirements, and the laser scanner specifications. The result was that the applications "Automatic emergency brake", "Stop & Go", "Precrash", "Pedestrian Protection", "Parking Assistant" and "Turning Assist" could all be supported by the following set of specifications:

Measurement range R	0,5 - 80 m
Measurement range on 5% reflectivity targets $R_{5\%}$	30 m
Distance measurement error σ	0.05 m
Scan frequency	10 to 40 Hz
Angular horizontal resolution α_{HOR}	0.25° to 1.0°
Scan angle	max. 240°
No. of scan planes (vertical layers)	4
Vertical field of view α_{VERT}	3.2°
No. of targets per shot	2
Laser class	1 (eye safe)

Tab. 1.   Technical data of ALASCA®.

As measurement principle of the laser range finder the "time-of-flight"-method was chosen. Here the travel time between the sensor and the target provides a measure for the distance of the target based on the fact that the speed of light is a constant.

Requirements such as the reliable detection of objects over of a wide horizontal angle and accurate longitudinal detection of the position are warranted by the specifications and the measurement principle.

In order to illustrate the way the specifications correspond to the applications the following example may be considered. A typical speed in an urban environment can be set to v_{typ} = 50 km/h. A dark target (i.e. pedestrian leg, black clothes) with a width of 13 cm and a reflectivity of 5% can be fully detected at $R_{5\%}$ = 30 m distance, owing to the accurate horizontal resolution of α_{hor} = 0,25°. The available response time, which corresponds to the velocity v_{typ}, is thus determined to be around 0.05 s to 0.06 s and is sufficiently large to activate safety measures.

During vertical pitching of the vehicle the ALASCA® is able to hold contact to the tracked objects because of the 4 scan-layers with a total angle of view of α_{vert} = 3.2°. Driving under rainy conditions or when the laser scanner window is contaminated by dust, are compensated by the ability of the opto-electronic detection scheme, which allows for the evaluation of two targets per laser shot. While the first "hit" on the laser path may be a raindrop, the second meaningful target stays on the tracked object.

The ALASCA® is fully integrated into the body of the test vehicle. For evaluation purposes 2 integration positions were chosen; vehicle front left/right, and front centre. It has to be mentioned that the integration of the scanner into the vehicle reduces the available horizontal scan angle. Depending on the desired application the integration position must be chosen respectively and some applications might require 2 laser scanners (i.e. front left and front right on vehicle) to ensure the full angular vision.

Classification of objects is necessary for many future applications and useful for pedestrian recognition.

ALASCA® classifies four groups of road user by a special algorithm:
- ▶ Trucks.
- ▶ Cars.
- ▶ Bicycles / Motorcycles.
- ▶ Pedestrians.

3 Automatic Emergency Braking (AEB)

3.1 Fundamentals

AEB is one of the ADAS applications that have been in the focus of the development within the current cooperation between HELLA and IBEO AS. Road accident statistics determined that about 50% of injuries caused to car passengers take place in head-on collision situations and about 40% of the drivers involved in these collisions do not brake at all. These figures reveal the potential of the AEB.

3.2 Functionality

The AEB function realized within the current project follows a straightforward strategy, which is laid out in figure 1. If there is an obstacle in the driving path, correct information about the location of the corners of the obstacle is necessary to calculate possible escape routes. At the point when the driver cannot avoid the collision anymore, the emergency braking is automatically activated. In the next moment the crash occurs. However, the consequences of the crash are greatly reduced compared to the situation where no or weak breaking is induced by the driver.

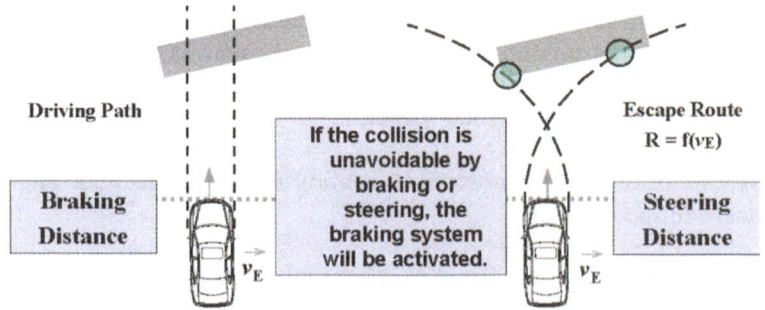

Fig. 1.   Schematic of the AEB situation. The obstacle is appearing in the vehicle's driving path (left). Contact with the obstacle cannot be avoided when the driving path associated with the minimum steering radius overlaps with the contour of the obstacle (right).

The system is on stand-by by starting the car. The AEB status is shown to the driver, like the airbag status. The driver does not take further notice of this system whilst driving, until an unavoidable accident occurs. On the occasion of an AEB event the driver will be informed by the HMI.

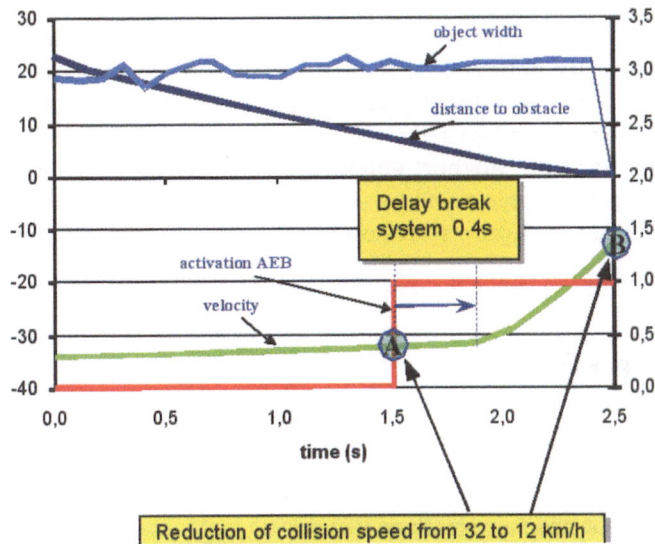

Fig. 2. Data traces from an experimental AEB situation as function of time. Note, that the vehicle velocity appears in negative numbers, starting at 32 km/h and ending at 12 km/h on contact with the obstacle. The AEB activation curve changes from 0 to 1 at 1.5 s, the point in time where the accident becomes imminent.

Figure 2 shows a set of experimental data taken during an AEB situation comparable to the one shown in figure 1. as a function of time. The road condition was dry and clean asphalt. The obstacle consisted of an arrangement of foam rubber blocks. Vehicle velocity, obstacle width, distance to the obstacle, and activation point of the AEB function are displayed. It can be seen that while the measured obstacle width remains nearly constant until the crash occurs (point B), the velocity (for reasons of convenience during programming shown in negative numbers) is strongly reduced after the activation of AEB as a step function (point A). The delay of 0.4 s compared to the electronic trigger set-off at 1.5 s results from the available electronic braking system, which had been developed for ACC and not optimised for AEB. Nonetheless a significant reduction of velocity from 32 km/h to 12 km/h is induced, which translates into a reduction of motion energy down to 14%.

3.3 Requirements on Brake Systems

The test vehicle used during this project is equipped with a standard series can bus system connecting all data sources and processing units. (Note: As an exception to this the ALASCA® transmits its data via a real time data bus to the processing unit on which the object recognition algorithm is implemented.) The brake booster consists of a modified ACC booster system. These two facts leave space for improvements in the future. While the reaction time of the driver can hardly be reduced, the deployment of a high power brake booster combined with a data bus system, which permits real time activation, will be very beneficial in further enhancing the effect of AEB. The next HELLA/IBEO AS test vehicle will be equipped with these features, which are also a requirement for future series production vehicles equipped with AEB.

4 ACC Stop & Go

4.1 Fundamentals

The application Stop & Go (S&G) is an enhanced function of ACC (Adaptive Cruise Control) for special situations like traffic jams. Like ACC it is a comfort function to reduce the stress of the driver but to let him have full responsibility for the vehicle.

ACC Systems are realized on RADAR and LIDAR basis. They are well adapted for far distance applications. For example the Hella ACC System IDIS® has a range of 150 m with a horizontal aperture angle of 16°. The detection area of an ACC System is different from an S&G system. S&G systems are well adapted for low speed and short-range targets. Therefore, it is the logical addition to the ACC function. For S&G, short and middle range distances in front of the vehicle are sensed with a large horizontal aperture angle (see figure 3). Also a high accuracy of the distance measurement data is necessary in cases of driving through small passages. Cutting-in of objects have to be recognized. These features are supported by the fusion of the ALASCA® and the IDIS® systems.

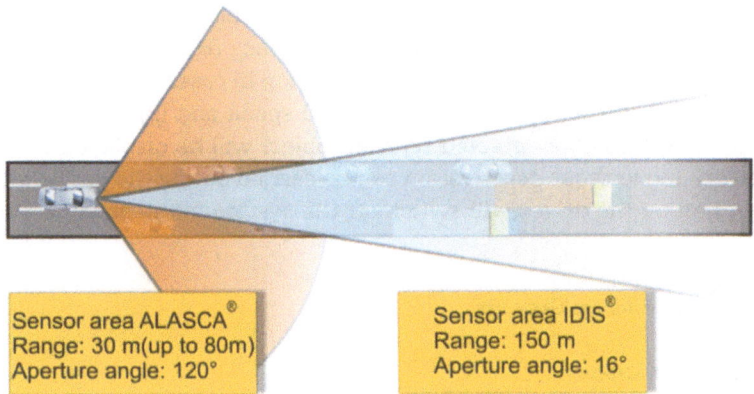

Sensor area ALASCA®
Range: 30 m(up to 80m)
Aperture angle: 120°

Sensor area IDIS®
Range: 150 m
Aperture angle: 16°

Fig. 3.   Overlapping field-of-vision areas of ALASCA® and IDIS®.

Extended requirements for the application ACC S&G compared to ACC only:

Large horizontal aperture angle	120° up to 150°
Fine distance measurement resolution	1 cm
Fine horizontal angular resolution	0.25° up to 1°
System availability for velocities	0 km/h up to 150 km/h and above
Considering of stationary obstacles	
Advanced deceleration values	

Tab. 2.   Extended requirements for ACC S&G

The current sensors for ACC, designed for long-range detection, and ACC S&G, designed for the short and medium range, cover all possible velocity ranges in the case of sensor fusion. For the sensor fusion, managing of the different modes of the respective sensors is necessary.

4.2   Driver Control

The driver is able to activate or deactivate the system. By activating the system he can select a desired speed and a desired following distance to the relevant object ahead. The ego-vehicle follows the relevant object ahead with a constant distance and stops automatically behind the relevant object if necessary. For a driver initiated or automatic restart ALASCA® scans the proximity due to its precise distance and angular resolution.

As the S&G system acts only with moderate deceleration, it is necessary to inform the driver via a Human Machine Interface (HMI) that he has to take over control of the vehicle via brake intervention in cases where collisions cannot be avoided. All set points for (1) selected speed and (2) selected following distance as well as (3) a detected relevant object will be displayed to the driver. Overdriving without deactivation can be carried out by the gas pedal. Brake intervention by the driver will switch off the system immediately.

4.3 Requirements on Sensors

4.3.1 Detection Range

The minimum detection range is defined by the covered distance for an automatic stop behind a standing vehicle. It is obviously defined within the limits of the maximum speed v_{max} and the maximum comfortable deceleration a_{max} of the driver's own vehicle. The tracking time for a relevant object and the response time given by the brake system define the total response time T_t. The minimum detection range d_{min} is defined by the following equation:

$$d_{min} = v_{max} T_t + \frac{1}{2} \frac{v_{max}^2}{a_{max}} + d_x \tag{1}$$

with d_x being a constant value describing a safety distance of 1.5 m.

The following table shows the minimum detection ranges for velocities of 30 km/h and 50 km/h assigned to comfortable deceleration values with an assumed response time $T_t=2$ s.

v_{max}	a_{max}			
	2 m/s²	3 m/s²	4 m/s²	5 m/s²
30 km/h	35.53 m	29.74 m	26.85 m	25.11 m
50 km/h	77.50 m	51.43 m	53.39 m	48.57 m

Tab. 3.   Table of minimum detection range for variable deceleration and different velocities.

A detection range of 30 m with a maximum speed v_{max}= 50km/h and maximum deceleration a_{max}= 5 m/s² implicates a response time of 0.66 s. This condition is fulfilled easily by the ALASCA® as shown in chapter 2. The following diagram shows the required detection range as a function of the response time calculated for a maximum velocity of 50 km/h. The ALASCA® system having a response time ≤ 0.5 s and a detection range of 30 m could be realized with the moderate deceleration value of a=2.5 m/s².

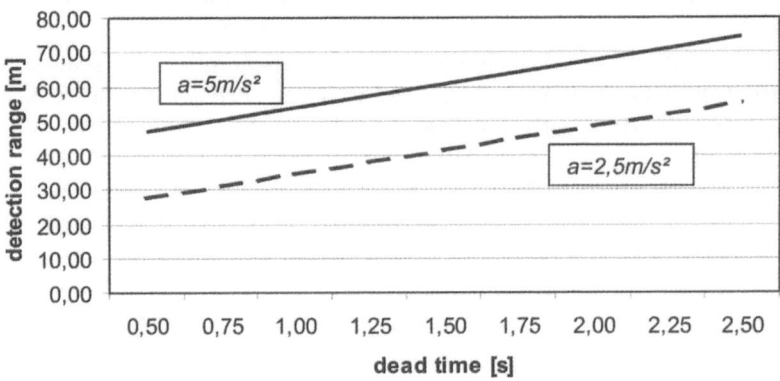

Fig. 4.   Detection range as a function of system response time.

The driver feels safe if the system notifies and acts on stationary objects quite early. At a velocity of 50 km/h and for a response time of 1.5 s the vehicle covers a distance of about 21 m before activating the system. At a velocity of 30 km/h this distance reduces to 12.5 m. For a presumed response time of 1.5 s a detection range of (1) 50-60 m for a velocity of 50 km/h and (2) 30-40 m for a velocity of 30 km/h is necessary.

In order to cover obstacles appearing during standstill of the vehicle, a detection range of 0.5 m is estimated as the minimum range.

4.3.2 Scan Angle

The scan angle α (figure 5) is defined to cover the ego lane and both adjacent lanes, in order to detect objects cutting-in:

$$\frac{\alpha}{2} = \arctan\left(\frac{X1}{X2}\right) \tag{2}$$

with X_1 being the lane width and X_2 being a comfortable stopping distance.

For a typical lane width of 3.5 m and a comfortable stopping distance of 1.5 m the half scan angle α is obtained to be 67°. By choosing this minimum scan angle it is ensured that the sensor detects cutting-in objects.

While driving through curves the sensor will hold contact with the relevant objects. For a scan angle of ±75°, curves with a radius ≥100 m are covered. ALASCA® outperforms this range with its theoretical maximum scan angle of 240°. In reality the scan angle is limited to a still sufficient value of 120° by the mounting position on the vehicle.

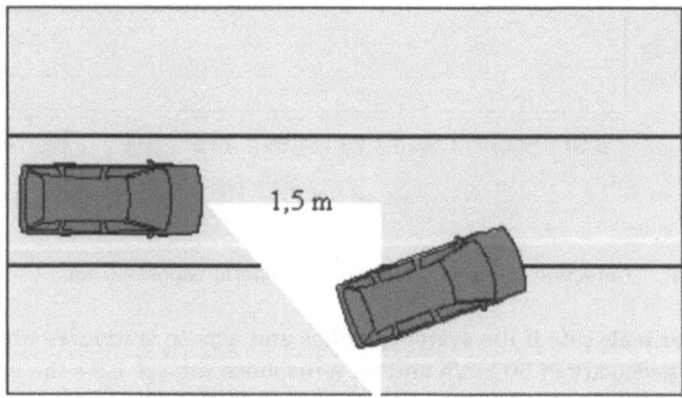

Fig. 5.   Illustration of the scan angle $\alpha/2$.

ALASCA® delivers an enhanced scan angle in comparison to common ACC Systems. Therefore in traffic jams during stand still ALASCA® monitors the proximity of the vehicle with high precision for automatic restart.

4.3.3 Vertical Field of View

Under S&G conditions enhanced drive dynamics are given compared to ACC. Enhanced limits for acceleration and deceleration, commonly +3 m/s² and −5 m/s² are considered. An increased vehicle pitch angle value of +0.8° / -2° results from this situation. ALASCA® is designed to cover this dynamic part of the vehicle pitch angle due to its vertical field of view of 3.2°.

The vertical field of view is divided into 4 levels offering a resolution of 0.8°. Therefore, it is also possible to distinguish between the road surface and the relevant object. During pitching of the vehicle due to road surface irregularities ALASCA® will continue to hold contact to the tracked objects.

4.3.4 Angular Horizontal Resolution

The new advanced task of the ACC S&G is to offer: (1) braking in front of stationary objects or (2) driving past parked vehicles at the lane side - demanding new requirements for the horizontal angular resolution. The requirements for the resolution are defined by the request of detecting the lane edge, the lateral position of the relevant object and the object width.

An example is shown in figure 6 for a typical situation with the ego vehicle in the centre lane and a parked car in the neighbouring lane. For passing the parked car on the right side a safety distance of $d_s \sim 30$ cm is demanded. The resolution of the y-coordinate should be at least half of the safety distance. For safe passing this requirement translates into an angular horizontal resolution of 0.3° in conjunction with a detection range of 30 m. ALASCA® fulfils the request with its horizontal angle resolution of 0.25° to 1°.

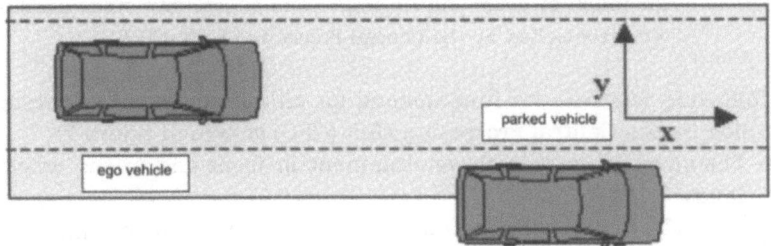

Fig. 6.   Illustration of the angular horizontal resolution.

4.3.5 Object Data

The following data of all relevant objects are delivered by the sensor system to enable sufficient brake and motor management.

Distance - The distance measurement resolution is 1 cm with a sigma of 5 cm.

Velocity - Velocity data are necessary for ACC-control and the detection of stopping and stationary objects, respectively. The resolution is 0.5 km/h minimum.

Acceleration - In the lower velocity range the acceleration of the relevant object is necessary for ACC control. Lateral acceleration values are necessary to detect cut-in and cutout manoeuvres.

Position - To select the relevant object for brake and motor management control, the object position in the (x, y)-coordinates of the right and left boundary, the (x, y)-coordinates of the object centre, and the object width are necessary.

4.4 System Concept

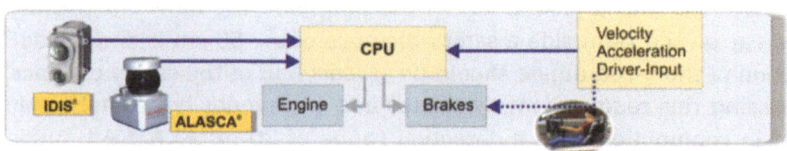

Fig. 7.   Basic elements of an autonomous S&G system containing the sensor fusion of IDIS® and ALASCA® and engine and brake management controlled by the Central Processing Unit (CPU).

The following features are fundamental for an autonomous S&G system and controlled by the Central Processing Unit (CPU) shown in figure 7:
  ▶ Scanning of the vehicle environment in front and in the neighboring lanes.
  ▶ ALASCA® calculates the object outline from the scan data and appropriate algorithms. The contour lines serve to extract and classify the objects. In addition, the distance, direction and relative velocity data of each object are supplied for further signal processing.
  ▶ Adopting of the relevant object from ACC-Software from IDIS® or tracking of relevant obstacles in the vehicle driving direction by ALASCA®. The object data are transmitted to the central processing unit (CPU). The CPU selects the relevant objects.
  ▶ Automatic adaptation of the speed setting to the distance and speed of the vehicle driving in front if necessary, up to a complete standstill. Based on the current data of the vehicle motion, the CPU directly controls the engine and the brakes.

▶ Auto restart under following conditions:
  ▶ All passengers are on board and vehicle doors are closed.
  ▶ No object is located between the driver's own starting vehicle and the
    tracked vehicle.

For safety reasons the driver is able to override the automatic control at ACC
S&G.

5   Additional Applications

5.1   Pedestrian Detection and Protection

In the future the number of fatal accidents of pedestrians will be reduced by
active and passive systems on the vehicle. Actively reacting systems require a
precisely measuring sensor like the ALASCA® in order to provide a reliable
database for introducing contour changes to the vehicle (lower bumper, active
hood). Therefore sensorial systems are needed that detect and classify pedestrians crossing the path of the ego vehicle. ALASCA® delivers high-resolution
range images in a wide horizontal angular range. Combined with advanced
sensor signal processing algorithms for high speed object detection, tracking
and classification this sensor provides the data required for future active
pedestrian protection systems.

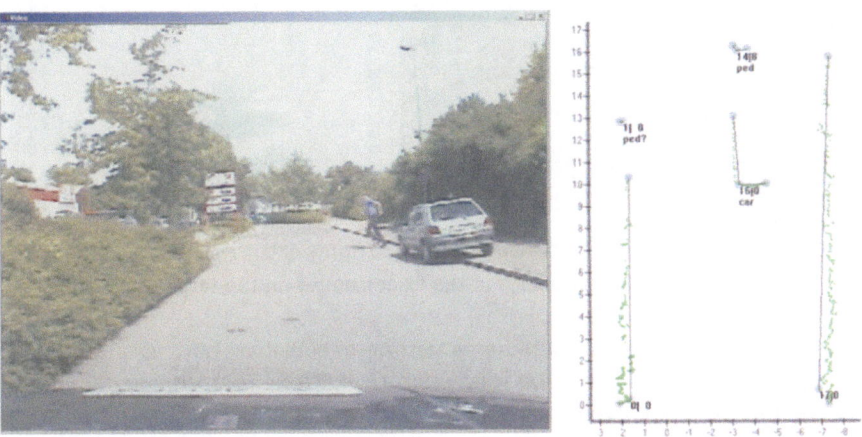

Fig. 8.   A video image (left image) and computed classification data (right
           image) of a scene showing a pedestrian just steeping onto the road
           from behind a parking car.

Pedestrians can be detected, tracked and classified even at larger distances as shown in figure 8, which allows their observation for a longer period of time in the scene. That allows a robust detection of potentially harmful situations.

5.2 Precrash

This safety system requires the detection and evaluation of an imminent collision. This system provides a larger time slot −compared to impact operated sensors− to activate safety measures like seat belt pre-tensioners, intelligent airbags, or the pretensioning of the brake booster. Reversible systems can be activated at a stage when a collision can still be avoided.

Upon the identification of an obstacle within the driving path the Time-to-Collision, Type of Collision Partner and Point of First Contact are calculated. Figure 9 shows a test scenario where the ego-vehicle is approaching an obstacle (box) of 40 cm width. In the video image, the scan data of the four ALAS-CA® scan planes are displayed in an overlay mode.

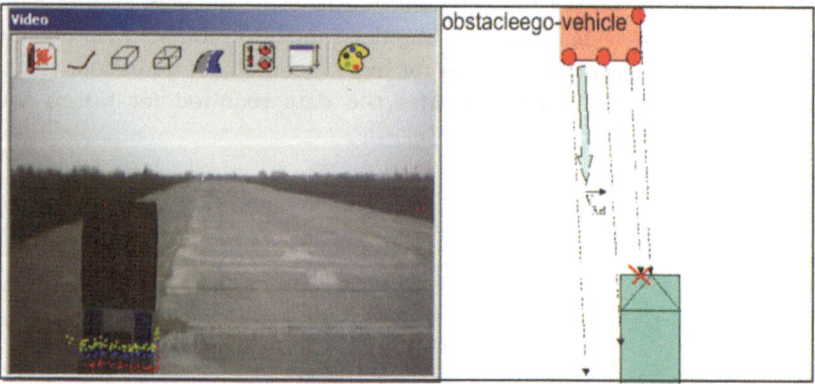

Fig. 9.   Video image (left) and schematic (right) of a Pre-Crash scenario.
Note the 4 scan planes of the laserscanner on the target-box.

Figure 10 shows crash-probability and time-to-collision, both as function of scans (on same scale). It can be seen from the graphs that while the number of scans increases the collision probability rises from 50% at scan #1 to 100% at scan #10 with a time-to-collision of approx. 100 ms remaining at the point of inevitable collision.

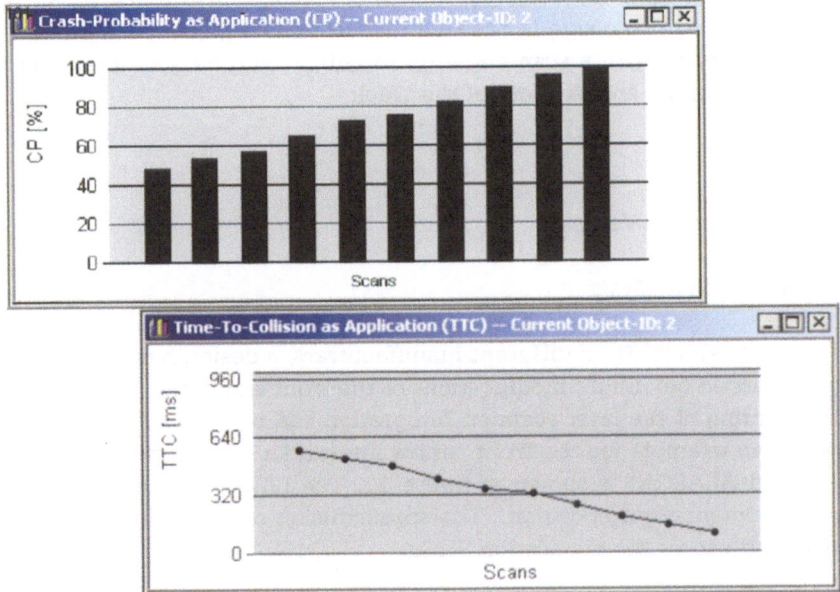

Fig. 10. Collision probability (top) and time-to-collision (bottom) of the Pre-
Crash scenario from figure 8.

5.3 Parking Assistant

While driving on a road with potential parking slots on the side these will be
evaluated and mapped by the parking assistant. ALASCA® measures with a
fine contour resolution the parking slot as well as different types of curbs. The
complexity of the parking assistant depends on the realization. A first step of
realizations might be a simple hint for a parking slot, like yes/no statements.
The next step might be automatic steering into the parking slot with brake
control of the driver. In future automatic parking will be realized.

5.4 Turning Assist for Trucks

Cyclists and people located in the blind spots of trucks are often overlooked by
the driver because of the lack of monitoring systems. This results in a large
number of fatal accidents. The implementation of a turning assistant will be an
important step to reduce the number of these accidents. It is realized by integrating an ALASCA® into the front right corner of the truck. The system monitors the area on right side of the truck with a scan angle of >90°. Thereby

vulnerable road users are detected. In case of a dangerous situation a warning signal (e.g. a red no-go sign in the side mirror) will be given to the driver. An advanced version with a scan angle up to 240° would be able to extend the field of view to the area in front of the truck.

6   Experimental Vehicles

6.1   Integration Analysis

For several vehicles from different manufacturers, a design integration analysis was carried out. Slight modifications of the front end of the car allow invisible mounting of the laser scanner. Integration can be realized in all types of cars. As an example for the front centre integration a car equipped with an integrated ALASCA® is shown in figure 11. The LIDAR system IDIS® is also shown in its mounting position. The sensor fusion of both systems is of particular interest for the application S&G.

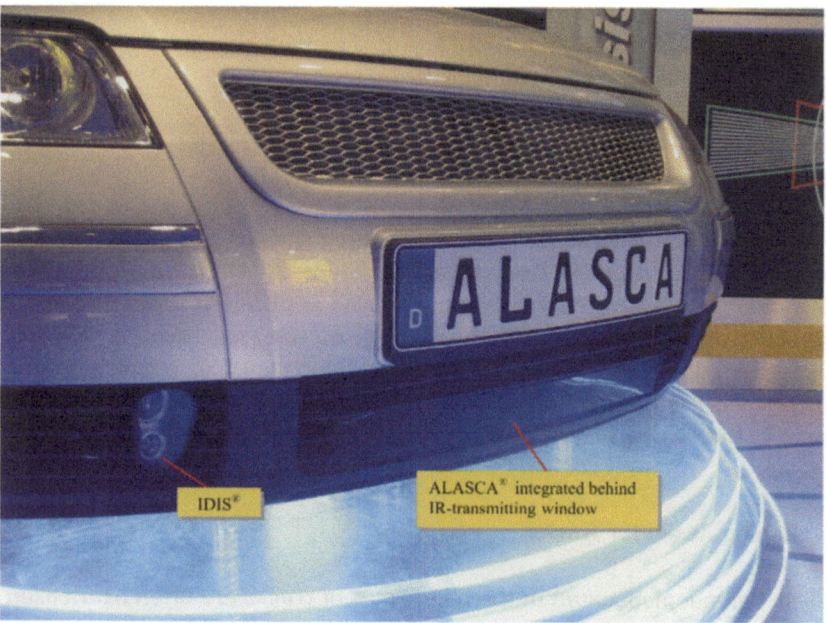

Fig. 11.   Integration of ALASCA® into the vehicle front end.

ALASCA® is mounted under the bumper in a centre position. For design reasons and protection ALASCA® is enclosed in a housing covered by an IR-transmitting window. This integration position permits a horizontal scan angle of up to 120°. The application S&G and AEB have been demonstrated with a test vehicle equipped with this integration solution.

6.2 Electronic Configuration

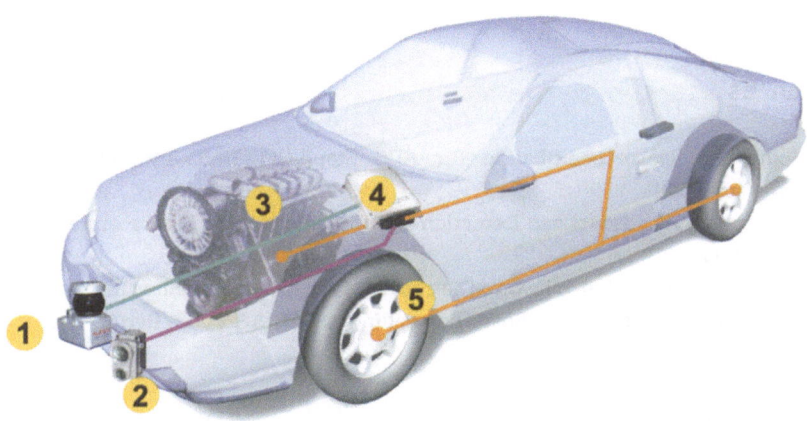

Fig. 12.   Vehicle architecture of sensors and data bus system. (1) ALASCA®, (2) IDIS®, (3) engine, (4) Central processing unit, (5) brakes.

The output signals of ALASCA® (1) are

- ▶ Availability judged upon self-test.
- ▶ AEB activation.
- ▶ Object.
- ▶ Object classification.
- ▶ Object velocity.
- ▶ Object position (x, y).
- ▶ Object acceleration.

These data are transmitted via CAN to the central processing unit CPU (4). The CPU also receives the object data generated by the ACC System (e.g. IDIS® (2)), the vehicle velocity and acceleration, etc. Depending on the application and the motor data, the CPU controls the engine (3) and the brakes (5).

The experimental vehicles are equipped with both applications, AEB and S&G.

7   Conclusion

In this paper an ADAS sensor technology in combination with its applications was presented. It was described in detail, how the laserscanner ALASCA® supports Automatic Emergency Brake, Stop and Go, Precrash, Parking Assistance, Pedestrian Detection and Turning Assist. In 2004 vehicles equipped with ALASCA® will be available for OEM-testing.

References

[1]   Lichte, B.: Sensorspezifikation für ACC Stop&GO. Hella-internal document, Lippstadt 2002

[2]   Fürstenberg, K.: Pedestrian Detection and Classification by Laserscanners. EAEC, Paris 2003

[3]   Lages, U.; Fuerstenberg, K. Ch.; Dietmayer, K. C. J: Laser scanner Innovations for Detection of Obstacles and Road. Proceedings of AMAA 2003, Berlin.

Brigitte Nitsche
Hella KG Hueck & Co.
Dept. GE-ADS
Beckumer Str. 130
59552 Lippstadt
brigitte.nitsche@hella.com
www.hella.com

Roland Schulz
IBEO AS GmbH
Fahrenkrön 125
22179 Hamburg
rsc@ibeo-as.de
www.alasca.info
www.ibeo-as.de

Keywords: Laser scanner, ACC Stop&Go, Automatic emergency braking

Sensor Fusion for Multiple Automotive Active Safety and Comfort Applications

N. Kaempchen, K. C. Fuerstenberg, A. G. Skibicki, K. C. J. Dietmayer,
University of Ulm

Abstract

In established driver assistance systems each application relies on its own sensor, which observes the vehicles environment. Advanced driver assistance systems (ADAS) have increasing demand for several sensor systems. The described laserscanner and video based sensor fusion approach serves as a general platform for multiple active safety and comfort applications. By fusing these sensors, a large field of view is obtained and the certainty and precision of the estimates in the relevant regions is increased significantly. Robust vehicle position measurements and classification, lane estimation as well as pedestrian recognition enable a broad support for applications such as lane departure warning, automatic emergency breaking, automatic cruise control for traffic jam situations (ACC Stop&Go), PreCrash, pedestrian protection and low speed following.

1   Introduction

Recent projects on driver assistance systems have focused on applications such as Pre-Crash (Chameleon) [1], ACC Stop&Go (Carsense) [2] and recognition of vulnerable road users (PROTECTOR). These advanced driver assistant systems (ADAS) have increasing demand for several sensor systems, which are not only complementary but also redundant. Some research has therefore been focused multi-sensor fusion [2-5]. The aim of such a step is to provide a fused description of the traffic scene surrounding the vehicle, which is relevant for ADAS, but not specific to a certain application. This fusion system incorporates the data of the diverse sensors into a single description. Thus the field of view of a single sensor is enlarged, the certainty and precision of the estimates is increased and additionally this system design is economically efficient, as different applications share a set of sensors.

In section 2 the motivation leading to the use of sensor fusion systems is explained and our approach specified. Section 3 deals with the laserscanners object tracking and classification as well as the pedestrian recognition. The image processing is explained in section 4. In section 5 some results of our sensor fusion system are presented and section 6 concludes the paper.

2   Sensor Fusion

2.1   Introduction

Established advanced driver assistant systems, such as Lane Departure Warning (LDW) and Automatic Cruise Control (ACC) have a common architecture, as shown in Figure 1a. The application is implemented for a special sensor configuration and therefore depends on these sensors. With an increasing number of applications which share a set of sensors, the sensor dependent part of the software would exist several times, i.e. found in each application (see figure 1b). Thus emerges a strong motivation to separate the sensor specific part of the system and use it as an interface between the sensors and the applications (see figure 1c). The sensor fusion module is responsible for collecting measurements, interpreting them sensor specifically and incorporating them into a unified, consistent description which is then forwarded to the applications. The sensor fusion must be in part sensor specific in order to obtain a maximal profit of each sensors measurement.

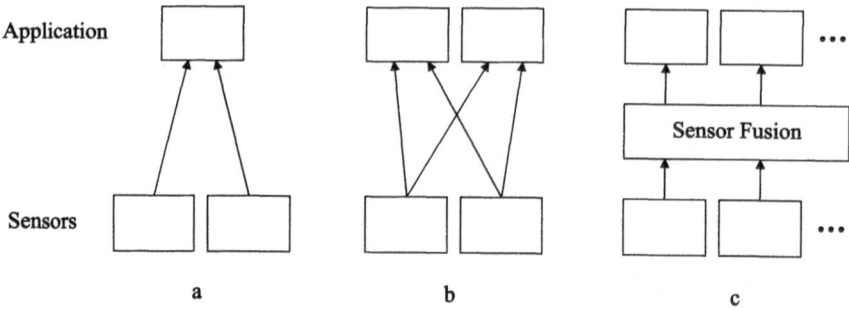

Fig. 1.   Driver assistant system architectures. (a) A single application which depends on one or several sensors. (b) Several applications each based directly on one or several sensors. (c) Possible future advanced driver assistant system architecture with common sensor fusion layer, thus separating the application from sensor specific implementations.

2.1 General High-Level Sensor Fusion

A general architecture of a centralised high-level sensor fusion system contains the basic components shown in figure 2 [6]. The environment description is composed of a set of objects, each of which is defined by an object state. There are different object models for trucks/busses, passenger cars, (motor) bikes, pedestrians and for stationary objects. Additionally the road is modelled as well as the ego-vehicle.

The objects of the environment description are predicted in order to synchronise them with the incoming sensor data. The object states are transformed into the parameter space of the sensor data by the use of inverse sensor models and are associated with the measurements. In case of an established association the sensor data is integrated into the object state. An object management handles the generation and deletion of objects in the environment model.

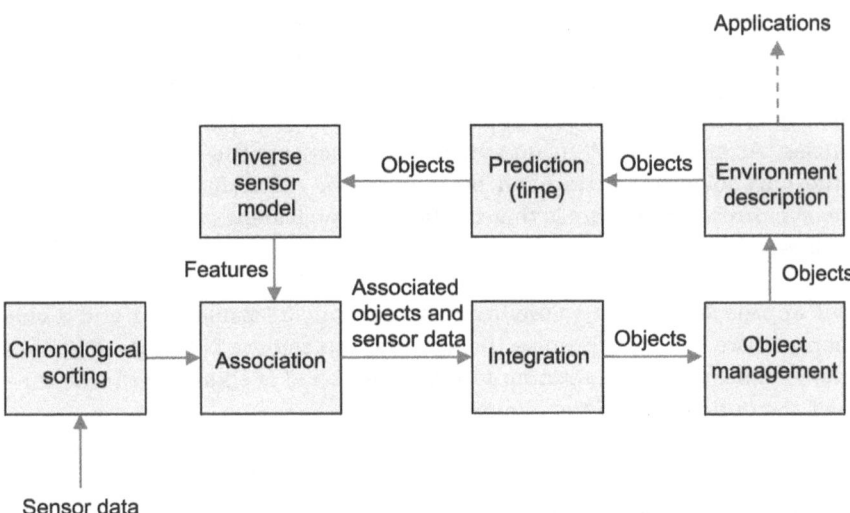

Fig. 2.   General architecture of a centralised high-level sensor fusion system which works as an interface between sensors and applications.

As mentioned, the sensor fusion depends on the sensor's characteristics. The sensor specific modules are the inverse sensor model and the association. These are the only components of the sensor fusion layer which need to have some knowledge of the sensor's characteristics. The integration and prediction may be performed using a Kalman-filter.

2.2 Synchronization

Sensor fusion combines several sensor measurements in order to obtain an enhanced object state estimation. The association and integration of sensor data requires its synchronisation with the environment description. The environment description is predicted for the time of measurement. Thereafter the sensor data can be associated and the object states updated. In order to obtain a precise synchronisation a sufficiently accurate global time for all sensors and the fusion system is necessary.

In order to obtain a time-consistent object state estimation the measurements have to be integrated in the order, they were acquired. This is a crucial constraint for a sensor fusion system. Depending on the synchronisation strategy, measurements have to be buffered for chronological sorting thus introducing a latency into the fusion system. For very time-critical ADAS such as PreCrash or automatic emergency brake, the overall system latency is of high interest. The advantage of such ADAS is drastically reduced if their reaction time is too long with respect to the high speed of passenger cars and therefore short time to contact in extreme situations. The aim is to minimise the systems latency which in our case is the latency introduced by the sensor fusion process. Different synchronisation strategies lead to varying worst-case sensor fusion latencies. As shown in [7] using synchronised sensors the latency introduced by the chronological sorting can be minimised. An additional advantage of using synchronised sensors is that the latency is constant over time and therefore determined.

In our approach we use two environment sensors, a laserscanner and a video camera. In order to synchronise the sensors, the camera is triggered at those points in time, when the rotating laserscanner head is aligned with the direction of the optical axis of the camera.

2.3 Calibration of Laserscanner and Camera

The sensors are calibrated in order to enable not only a temporal alignment given by the synchronisation but also a spatial alignment. Figure 3 shows an image where the laserscanner data is projected into the image domain. By means of an accurate synchronisation and calibration, image regions can be associated directly with laserscanner measurements. Therefore it is possible to assign certain image parts a distance, which is a major advantage of this fusion approach. In figure 3 the laserscanner measurements on the back of the preceding car fit very well with the image taken by the camera. The precision of the association between the scans and the image can also be seen on the bollards delimiting the area of the road works.

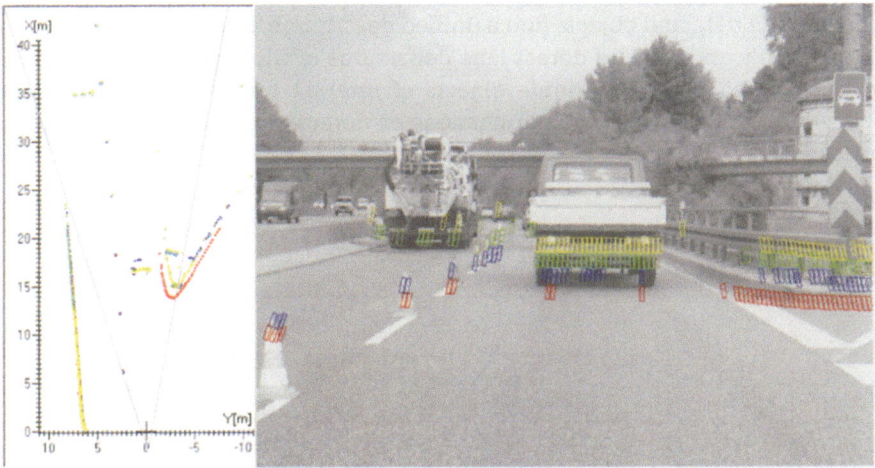

Fig. 3.   Laserscanner measurement in the bird view. The different layers of the multilayer laserscanner are colour coded. In grey the field of view of the camera (left). Laserscanner data projected into the image domain (right).

2.4   System Architecture

The sensors used for our sensor fusion approach are a laserscanner with four scan planes, a monochrome monocular camera and an Inertial Navigation System (INS). The laserscanner is mounted at the front bumper of our test vehicle and the camera behind the front screen beside the inner mirror. The field of view of the laserscanner compared to the field of view of the camera is shown in figure 3. The horizontal view of the laserscanner is with 180° much larger than that of the camera with 28°. However, the horizontal resolution of the laserscanner is with 0.25° (0.5°) at 10 Hz (20 Hz) much lower than the resolution of the 640 x 480 pixels image with 0.045°.

Both sensors are combined in the fusion system in a way in order to enable a high amount of synergetic effects. The laserscanner tracks and classifies vehicles, motorbikes and pedestrians in the 180 degree environment in front of the ego vehicle. The image processing unit estimates the lane and the position of the ego vehicle in the lane. Another image processing module refines the lateral position and velocity estimation of the objects tracked by the laserscanner. At times when the objects exceeds the object detection distance of the laserscanner the object can be tracked for a while in the image.

The high-level fusion combines the ego motion estimation, the lane recognition and the detected objects into a unified description. Based on the high-level fusion, applications could detect lane departures of the ego vehicle, obstacles in the driving path, determine objects of interest for ACC and Low Speed Following, as well as their lane changes or detected potentially endangered pedestrians.

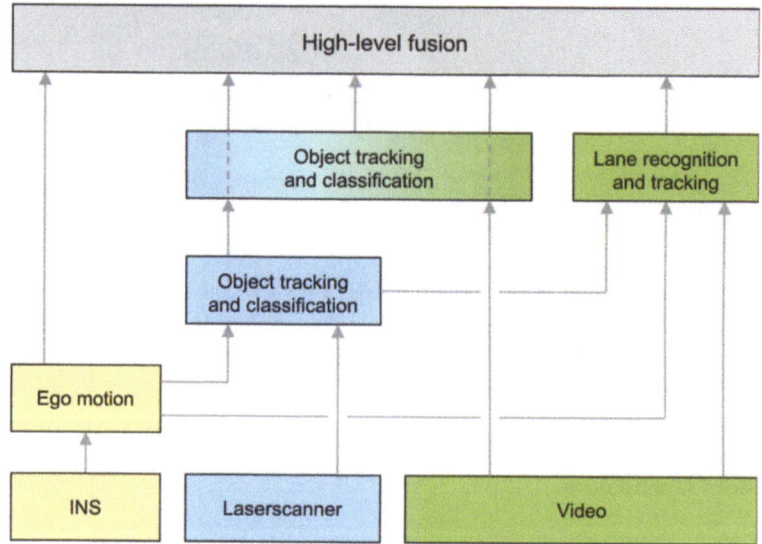

Fig. 4.   System architecture involving different levels of sensor fusion.

3   Laserscanner

There is a need for sensorial systems which are able to observe the vehicles environment in order to detect and classify endangered pedestrians and to track vehicles. laserscanners deliver high resolution range images in a wide horizontal angular range. Combined with advanced sensor signal processing algorithms for high speed object detection, tracking and classification, these sensors are able to provide the sensorial data required for future active safety systems.

The multilayer laserscanner (LD ML) used for environment observation uses 4 scan planes instead of one which allows for a compensation of the pitch angle of the ego-vehicle [8].

Fig. 5.   Multilayer laserscanner LD ML.

3.1 Object Detection

The distance measurements generated by one revolution of the rotating head (scan) are divided into segments, which are assumed to belong to the same object. The contour of a segment representing a vehicle consists typically of one or two straight lines. The appearance depends on the orientation and distance of the object in the laserscanner's co-ordinate system, as shown in figure 6. Pedestrians typically appear as a small cluster of measurement points.

3.2 Object Tracking

In order to track objects close to the ego vehicle the predicted object states of these objects have to be associated with current measurements (segments).

In general the estimation of the ego-motion under defined driving conditions is based on the vehicle's INS-Data [8]. A reliable ego-motion estimation under both, defined and undefined driving conditions, can be determined by using the relative velocity of static objects in the vehicle's environment such as posts, reflection posts and trees [9]. The ego-motion of the test vehicle is compensated before segments are assigned to object tracks in order to obtain a more reliable and robust tracking. The typical Kalman filter cycle is performed in three steps.

First, the state vector $\mathbf{x}$ and the state error covariance matrix $\mathbf{P}(k)$ are pre-dicted and the ego motion compensation is performed. The object model used is a single-point-mass model, which simply represents an object as a mass that can move in the 2D space without any cinematic constraints. For each model class, a constant velocity is assumed. Therefore accelerations of the objects are considered in the model covariance. Different acceleration limits depending on the object's class lead to the definition of class specific covariances. Therefore accurate gating areas are obtained, as shown in figure 7. For more details see [10, 11].

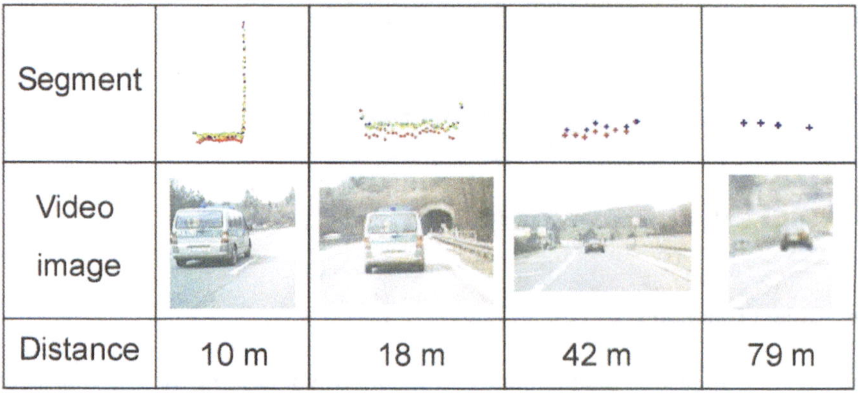

Segment				
Video image				
Distance	10 m	18 m	42 m	79 m

Fig. 6. Segments with associated video image sorted by distance. The segment contour and the number of scan layers (coded in different colours) depends on the distance.

Second, segments are assigned to object tracks, which will be discussed in the next section.

In the third step a new measurement (object reference point) is incorporated into the a priori estimate of the Kalman filter, thus improving the a posteriori estimate.

3.4 Object Association

The data association is responsible for the assignment of segments to object tracks. The association process can be divided into three sub-processing stages. The computation of the gate, checking of the gate criteria and the assignment of segments to the objects.

First, gating is used as a screening mechanism in order to determine which segments are valid candidates to be associated with an object track. Gating is performed primarily to reduce unnecessary computation by the association algorithm. A rectangular gate is used whose orientation is determined by the absolute velocity vector of the object. The size of the rectangular gate is adapted to the vehicles dimensions (width w_i and length l_i) and to the quality of the state estimation (figure 7).

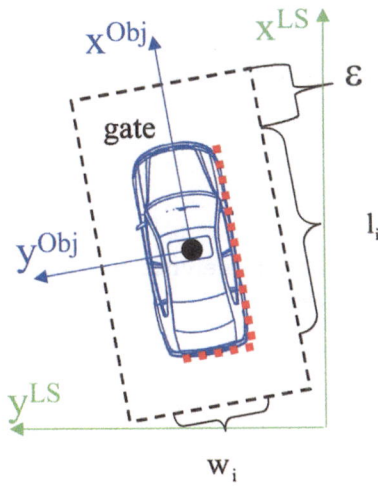

Fig. 7.   Gate of the association step to assign the matching segment.

If several segments meet the gate criterion, an object reference point is calculated for each segment in order to determine the closest segment or segment combination. Under real world conditions, it can be difficult to accurately identify an object reference point, because of the changing contour and conditions of the segments, as shown in figure 9.

In order to determine the object reference point three different methods are implemented and evaluated, which will be discussed in the following. They are compared and evaluated based on simulated data, which enables an objective comparison.

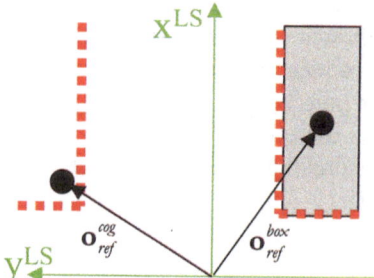

Fig. 8. Object reference point calculated by the centre of gravity approach (left) and the two line approach (right).

Centre of Gravity Approach

This approach calculates the object reference point as the mean of all segment points, as shown in figure 8 (left):

$$\mathbf{o}_{ref}^{cog}(x,y) = \frac{1}{n_{point}} \sum_{i=1}^{n_{point}} p_i(x,y) \tag{1}$$

Where n_{point} is the number of points of a segment and $p_i(x,y)$ is the position of a point in the laserscanner's co-ordinate system. This strategy considers all points of a segment for the computation of the object reference point and is used for all objects, except for vehicles. Considering a passing vehicle the object reference point moves inside the object contours, as shown in figure 9, which would lead to inaccurate velocity estimates.

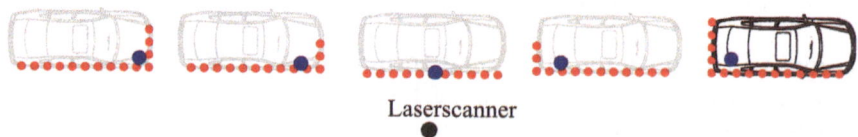

Laserscanner

Fig. 9. A car passing in front of the ego-vehicle, which is equipped with a laserscanner. The blue dots represent the object reference point, which is calculated by the centre of gravity approach.

Object-Box Approach

This approach is designed to track vehicles. Vehicles are represented by a rectangular box with adaptive length and width, the so called adaptive object box, which is calculated from the segment's parameters.

The points of a segment are sorted clockwise by their angle, where the first point $\mathbf{f}$ of a segment is the point with the smallest angle and the last point $\mathbf{l}$ is the segment-point with the largest angle. The measurement point of the segment with the shortest distance to the scanner c is called the closest point. A straight-line-object is represented by a segment with straight line contour (see figure 6, distance 18 m, 42 m and 79 m and figure 10). In consequence, an angle-object is represented by a segment with angular shape (see figure 6, distance 10 m).

Two strategies were developed to compute the rectangular box. It is differentiated between objects of known velocity and objects without known velocity. Moreover, it is distinguished between straight-line-objects and angle-objects, especially for vehicles like cars and trucks.

Straight-line-objects

The heading of a moving object is determined by the velocity vector. Comparing the heading with the vector from the first to the last point $\mathbf{u}_{FL}$ of a segment, the length and width of the tracked object can be calculated. Using a priori knowledge the adaptive object box with the actual length or width is computed.

By means of the normal vector

$$
\mathbf{n}_{FL} = \frac{1}{\sqrt{u_{FL_y}^2 + u_{FL_x}^2}} \begin{pmatrix} -u_{FL_y} \\ u_{FL_x} \end{pmatrix} \tag{2}
$$

the object reference point for vehicles moving radial to the laserscanner is computed by

$$\mathbf{o}_{ref}^{box} = \frac{\mathbf{f}+\mathbf{1}}{2} + \frac{O_{length}}{2} \cdot \mathbf{n}_{FL} \tag{3}$$

If the vehicles are moving tangential to the laserscanner, the object reference point is given by

$$\mathbf{o}_{ref}^{box} = \frac{\mathbf{f}+\mathbf{1}}{2} + \frac{O_{width}}{2} \cdot \mathbf{n}_{FL} \tag{4}$$

where O_{length}, O_{width} represent the length and width of the object, respectively.

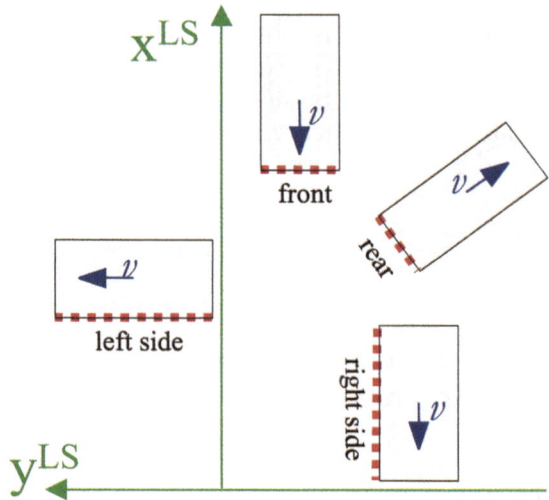

Fig. 10. Different representation of a vehicle by a segment, depending on the angle of view to the object.

Straight-line-objects without any velocity are always represented by a segment which corresponds to the length of the tracked object. The object box will be aligned on the segment with the actual length. If a priori knowledge is available the object box has the maximum width of the tracked object that ever occurred. The object reference point is the centre of gravity.

Angle-objects

For moving objects the heading of the object can be calculated from the velocity vector. Comparing the heading with the vector from the first to the closest point $\mathbf{u}_{FC}$ and the vector from the last to the closest point $\mathbf{u}_{LC}$, it can be determined which line of the segment represents the width and which line the length of the tracked object. Using the vector from the first to the last point of a segment $\mathbf{u}_{FL}$ and the velocity vector $\mathbf{v}$ as shown in Figure 11, the angle between both vectors can be computed and thus the length and width of the object.

Fig. 11. Computation of length and width of the vehicle using a) the velocity aligned vector $\mathbf{u}_{FL}$ (left) b) without known velocity, the vectors between the first, last and closest points (right).

Knowing the length and width of the object, the computation of the object reference point can be easily done. As shown in the following equations it has to be differentiated which line of the angle-object is parallel to the velocity vector $\mathbf{v}$.

$$\mathbf{u}_{FC} \parallel \mathbf{v}: \qquad \mathbf{o}_{ref}^{box} = \mathbf{f} + \frac{d_{length}}{2} \cdot \frac{1}{|\mathbf{u}_{FC}|} \mathbf{u}_{FC} + \frac{O_{width}}{2} \cdot \mathbf{n}_{FC} \qquad (5)$$

$$\mathbf{u}_{LC} \parallel \mathbf{v}: \qquad \mathbf{o}_{ref}^{box} = \mathbf{l} + \frac{d_{length}}{2} \cdot \frac{1}{|\mathbf{u}_{LC}|} \mathbf{u}_{LC} + \frac{O_{width}}{2} \cdot \mathbf{n}_{LC} \qquad (6)$$

In the case of a non moving object, it depends on the object class which method is used to compute the object reference point. Objects that represent vehicles can be described by an object box even without velocity. However, it is not possible to determine a reliable length and width of the object, therefore the longer line is supposed to be the length and hence the shorter line to be the width. The object reference point can be computed by means of the vectors $\mathbf{u}_{FC}$ and $\mathbf{u}_{LC}$ which are supposed to be almost orthogonal to each other.

Two Line Approach

The two line approach uses one Kalman filter (L) to track the length and one Kalman filter (W) to track the width of the object. As shown in figure 12, the measurement points of a segment are divided into two straight lines. The division of the segment is done analogous to the width and length computation in the previous section.

Using the two line approach, also two object reference points are required to update the Kalman filter (L, W). In this case the object reference points are the geometrical centres of the lines. The motivation for this approach is that depending on the orientation in the laserscanner co-ordinate system, the segments consist of either one or two straight lines, as shown in figure 12. If there is only one straight line, just the appropriate Kalman filter will be updated. Whenever both straight lines, a so called angle-object is detected, both Kalman filter are updated to track the object.

To compute the velocity of an object, both Kalman filter elements for width and length are weighted depending on their performance. The weighting is based on the innovation vector $\acute{a}(k)$.

$$\acute{a}(k) = y(k) - C \cdot x^-(k \mid k-1) \tag{7}$$

where $y(k)$ is the temporary generated measurement vector and $x^-(k|k-1)$ is the predicted state vector of the previous step. This leads to a weighting factor $g(k)$

$$g^{W/L}(k) = \acute{a}_x^{W/L}(k)^2 + \acute{a}_y^{W/L}(k)^2, \quad g(k) = \frac{g^W(k)}{g^W(k) + g^L(k)} \tag{8}$$

Consequently, the computation of the velocity vector v is done by weighting both the velocity vector of the width, v^W, and the length, v^L, with the factor $g(k)$ as follows

$$v(k) = g(k) \cdot v^W(k) + (1 - g(k)) \cdot v^L(k) \tag{9}$$

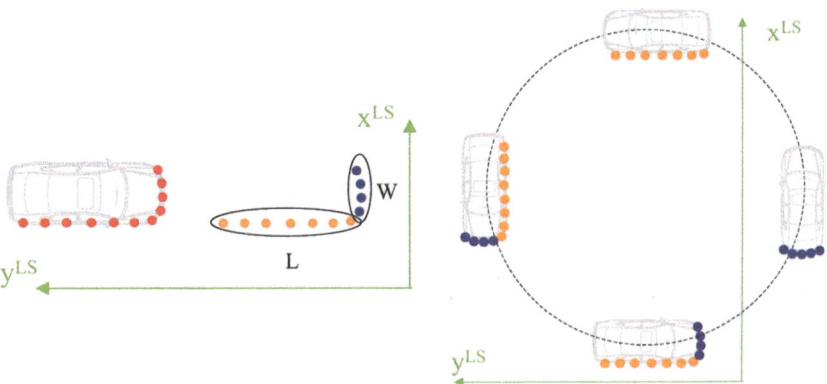

Fig. 12. The measurements at a vehicle are divided in two straight lines (left). A vehicle can be represented by one or two straight lines, depending on the angle of view (right).

To handle the changing of the segment contour, the change of the size (length or width) of an object is considered. As illustrated in figure 13, the real movement of the object displayed with a blue arrow and the movement of the object reference point, represented by black arrows, differ depending on the angle of view.

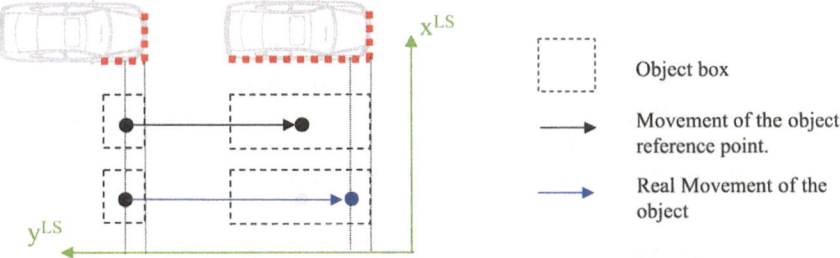

Fig. 13. Change of the size (here: length) of an object due to different angles of view which causes a change of the object reference point.

In this case an imaginary movement of the object reference point would occur, which is a result of a change of size of the current object. To avoid the imaginary movement of the object, an object reference point is calculated, which is at the same point in the object co-ordinate system as in the previous scan.

The computation of the object reference point of the innovation $\mathbf{o}^i_{ref}$ consists of the actual object-length d_{length}, the maximum occurred object-length O_{length} and the vector $\mathbf{u}_{FC}$.

$$\mathbf{u}_{FC} \parallel \mathbf{v}: \qquad \mathbf{o}^i_{ref} = \mathbf{f} + (d_{length} - \frac{O_{length}}{2}) \cdot \frac{1}{|\mathbf{u}_{FC}|} \mathbf{u}_{FC} \tag{10}$$

$$\mathbf{u}_{LC} \parallel \mathbf{v}: \qquad \mathbf{o}^i_{ref} = \mathbf{f} + (d_{length} - \frac{O_{length}}{2}) \cdot \frac{1}{|\mathbf{u}_{LC}|} \mathbf{u}_{LC} \tag{11}$$

After the innovation the real object reference point $\mathbf{o}^L_{ref}$ is set into the state vector of the Kalman filter to obtain a more reliable prediction in the next Kalman filter cycle.

$$\mathbf{u}_{FC} \parallel \mathbf{v}: \qquad \mathbf{o}^L_{ref} = \mathbf{l} + \frac{d_{length}}{2} \cdot \frac{1}{|\mathbf{u}_{FC}|} \mathbf{u}_{FC} \tag{12}$$

$$\mathbf{u}_{LC} \parallel \mathbf{v}: \qquad \mathbf{o}^L_{ref} = \mathbf{l} + \frac{d_{length}}{2} \cdot \frac{1}{|\mathbf{u}_{LC}|} \mathbf{u}_{LC} \tag{13}$$

The introduced methods are compared and evaluated based on a scenario with simulated data, where a car passes the laserscanner, as displayed in figure 9. The velocity of the car, calculated by the tracking algorithms, is a good indicator for the quality of the different tracking techniques. Therefore the obtained velocities of the different approaches are compared with the predetermined velocity of the simulated object, as shown in table 1.

Method	Standard deviation 3 σ
Centre of Gravity approach	7.7 km/h
Object box approach	1.1 km/h
Two Line approach	0.3 km/h

Table 1. Evaluation of the different methods with simulated laserscanner data for a passing car.

However the evaluation on simulated data is only the first step in the development of new algorithms. Ongoing evaluations on real data show that real vehicles are tracked very precise as well. As in real scenarios the true velocity is not known, estimations are done to compare these with the calculated velocity of the object. Without doubt the calculated velocity of vehicles based on real data would not be as good as the simulated results, but the estimated accuracy of about ±1 km/h in first investigations with the two line approach is convincing anyway.

Segment to Object Assignment

There are several well known procedures to assign segments to objects. The usual assignment method is the global nearest neighbour method. However, there are more sophisticated methods for the assignment problem.

In the current algorithm an assignment matrix, as shown in figure 14, is used in conjunction with the Hungarian method for the assignment problem. The columns of the assignment matrix represent the objects and the rows represent the segments. Every time a segment is within an object-gate, there is an entry in the assignment matrix at the equivalent position.

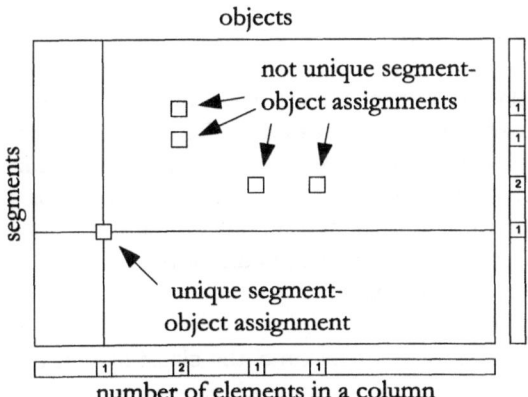

Fig. 14. Assignment matrix to combine the matching segments to the object tracks.

Unique segment-object assignments can be obtained in the assignment matrix and an instantaneous assignment can be carried out. This is the case, if only one segment is within a gate. These unique assignments are cancelled in the assignment matrix and a sub-matrix with only non unique assignments is generated.

The sub-matrix with non unique assignments has to be further processed. For that we use the Hungarian method, which returns the best possible assignment. The Hungarian method solves all non unique segment-object assignments.

3.5 Object Classification

The object classification in general is based on typical a priori known object-outlines (static data) of typical road users, such as cars, trucks/busses, motorcycles/bicycles and pedestrians. Additionally, the history of objects classified earlier and the dynamic state variables of the tracked objects are taken into account to improve the classification performance and its reliability.

In case there is not enough information for an object classification a hypothesis is generated, based on the objects current appearance. The temporary assignment is valid, as long as there is no violation of a limiting parameter. However the classification is checked every scan to verify the assignment of the specified class.

3.6 Pedestrian Recognition

The strategy to classify pedestrians is based on the following idea: An object that looks like a pedestrian with respect to its outline is classified as a pedestrian only, if the object is presently moving or has moved during the period of tracking in the past. Otherwise it will be classified as a non moving obstacle. However, if the non moving pedestrian, classified and tracked as a non moving obstacle, is located in the predicted path of the ego-vehicle, the situation is rated to be dangerous anyway.

If an object is detected for the first time, the classification result is based solely on its outline. As in this case a small object can be either a pedestrian or a small static object, such as a pole or a lamppost, the classification result is 'ped?'. As long as a small object is not moving towards the road, there is no hazard implied, neither for the vehicle nor the object itself. If a small object is moving, it is immediately classified as a pedestrian, based on the previous initial classification 'ped?' and its absolute velocity [12].

Slow pedestrians, which are able to change their moving direction in a split second, are tracked with the simple single-point-mass model, with the same limiting parameters to all directions of movement. If the pedestrian is moving faster (> 20 km/h), it can neither change the moving direction nor increase the speed rapidly, just deceleration could be faster. For that reason different longitudinal and lateral limiting parameters are used to model this behaviour. A running pedestrian has a similar model as a bike with the same speed.

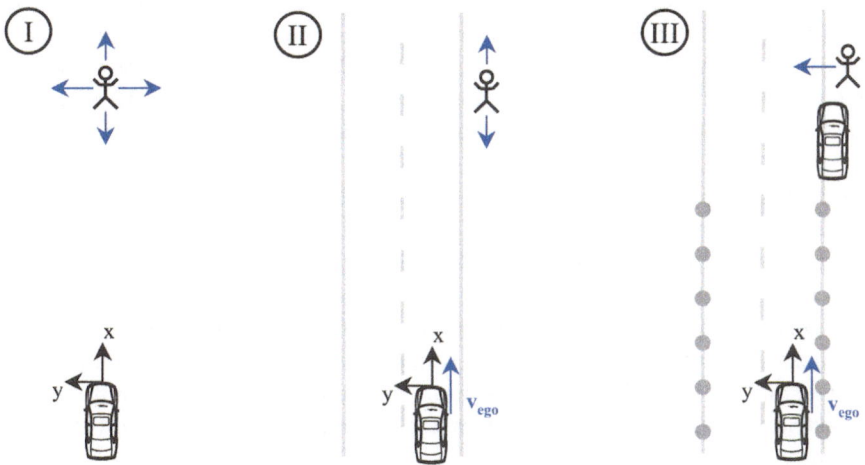

Fig. 15. Overview on the different test scenarios.

The pedestrian recognition algorithm based on range images of a laserscanner was investigated in different test scenarios. It is important that the system detects and classifies every pedestrian in the region of interest (ROI) at a high confidence level. The performance is assessed by the 'true positive' rate. The obstacles, which are classified wrongly as a pedestrian are counted resulting in a 'false positive' rate.

In scenario I the ability to detect and classify pedestrians at different distances and angles was tested (figure 15). Here, a pedestrian walked in an unstructured environment (free parking space). The ROI was defined with a longitudinal range from 0 m to 60 m and a lateral range of ±8 m in front of the ego vehicle. The tests performed in scenario I, which comprises more than 3000 scans in total, results in a true positive rate of 100%. The false positive rate was 0.0. This is not really surprising as no other objects existed within this test.

Scenario II was performed in an urban environment. The ego vehicle was driving with different velocities of up to 50 km/h. A pedestrian is walking along the road. The ROI is defined in a longitudinal range from 0 m to 60 m and a lateral range of ±5 m in front of the ego vehicle. The pedestrian walking along the road were detected and classified all the time in the ROI (true positive rate of 100%). Sometimes other objects were classified as a pedestrian at the boundary beside the street (false positive rate of 20%). By introducing a filter that suppresses the output until the object was classified 3 times within the same class a false positive rate of 0% can be achieved even in this scenario. However, due to this filter process, potential pedestrians are output to the application with a delay. The total delay depends on the scan frequency of the laserscanner. At 10 Hz scan frequency this delay would be about 300 ms, using 40 Hz scan frequency the delay would be about 75 ms. However the update rate after initial classification is 100 ms, 25 ms respectively.

The last scenario III was carried out on a test field, where tubes mark the boundary of the road, as shown in figure 15. A pedestrian is stepping onto the marked road initially occluded by a parking car. The classification of the pedestrian crossing the street was always done correctly in this test case (true positive rate of 100%). No other objects than pedestrians were detected belonging to the class pedestrian (false positive rate of 0%). However, two scans are needed to compute the velocity and perform the first classification result after the pedestrian enters the field of view of the laserscanner from behind the parking car.

	True Positive	False Positive	False Positive (with additional filter)
Scenario I	1.0	0.00	0.00
Scenario II	1.0	0.20	0.00
Scenario III	1.0	0.00	0.00

Table 2. Summary of the classification results.

The classification results of the first approach to recognise pedestrians as discussed above are summarised in table 2. They illustrate a good performance for less structured environment. However, there are still scenarios which will be a challenge for pedestrian recognition, e.g. highly frequented urban scenarios with lots of pedestrians walking close together. Also the differentiation of a post or a tree and a non moving pedestrian in a range image of a laserscanner can not be done up to now.

4 Video

One advantage of the laserscanner measurement principle is, that only there, where the laser beam is reflected by an object, a measurement is performed. Therefore the laserscanner is very reliable in object detection. Additionally the distance is measured very accurately. Both characteristics are weak points in monocular image processing approaches. On the other hand a video image exhibits a higher angular resolution than a laserscan. Therefore we take the advantage of both systems and design an image processing algorithm which is attention driven by the laserscanner's object detection and which refines the lateral position and velocity as well as the width measurements.

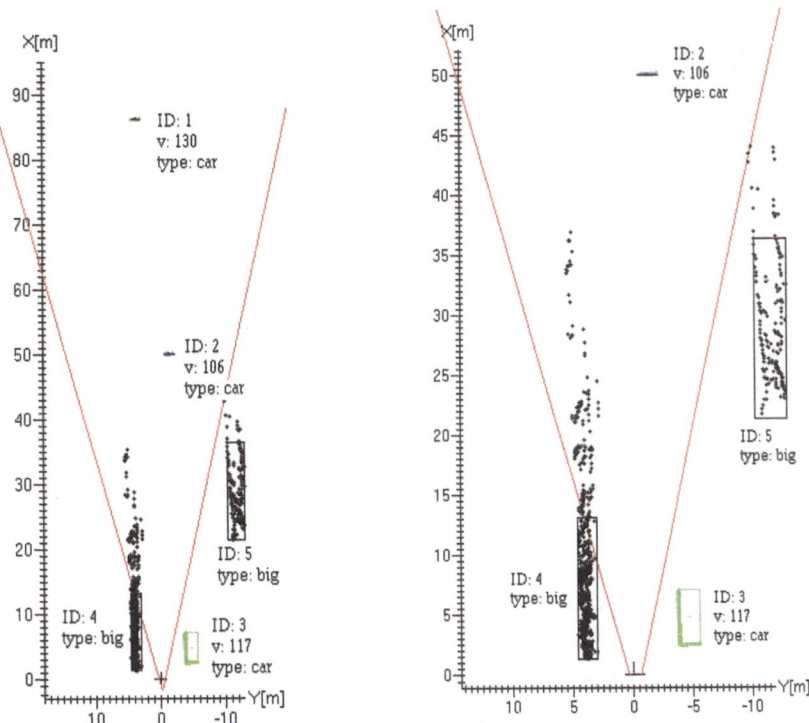

Fig. 16. Bird view of a typical scenario on a highway. In black the laser-scanner measurements. Three cars are detected by the laserscan-ner. In red the field of view of the camera.

4.1 Region of Interest

In figure 16a typical highway scenario is shown. Three cars are detected by the laserscanner. Two cars are in the overlapping field of view of both sensors and one car is only seen by the laserscanner. The objects detected by the laserscanner are projected into the image domain via a homogenous transformation given by the calibration. No temporal alignment is necessary, as the image acquisition is synchronised with the laserscanners measurement cycle.

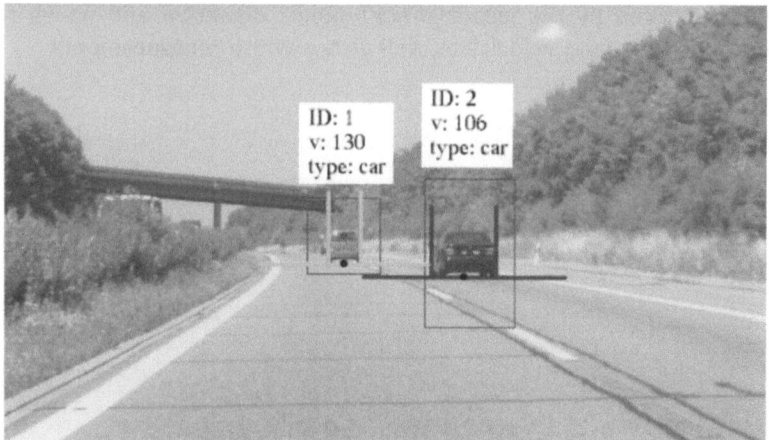

Fig. 17. In violet the ROIs given by the laserscanner. The U-shapes show the width and position of the cars refined by the image processing.

A region of interest (ROI) in the image is calculated from the objects width, measured by the laserscanner and the type of the object, classified by the laserscanners object recognition (see figure 17).

Object Recognition

The ROI of each object is searched for features which exhibit the presence of a vehicle. Depending on the classification of the laserscanners object recognition into cars and trucks, different features are searched for. Using adaptable thresholds the applied image processing is widely independent of illumination changes.

Shadow

A first feature is the shaded underbody of vehicles. In day time scenarios this shadow is almost independent of the illumination. For cars in the far field (>50 m) this dark area takes a special shape. As shown in figure 18a the wheels and a few image rows above the wheels exhibit dark grey values.

(a)   (b)   (c)

Fig. 18.   ROI of image (a, left). In green the wheel pattern hypothesis in the
           binary image (b, 2nd from left). Vertical, horizontal and combined
           edges in the image (c, three pictures at right hand side).

In order to search for this pattern of dark area, a binary image is calculated from the ROI. In figure 18b all pixels which are below a threshold are shown in orange, the others are drawn in black. Possible shapes and sizes of the template's appearances in the image domain are motivated by the possible width of passenger cars as well as the real sizes of their wheels. The templates are scaled with respect to the distance of the vehicle measured by the laserscanner. Therefore the wheel pattern is searched for in every row of the ROI taking minimal and maximal possible parameters of width of cars and their wheels into account. Every match is taken as a hypothesis. In figure 18b the green horizontal lines show the regions a matching wheel template was found. Then the ROI is searched for shadow regions which are above the hypothesis of the wheel pattern and stem from the dark areas beneath cars. For all hypothesis combinations a probability is calculated. If the probability of the best hypothesis exceeds a threshold it is taken as a match for the recognition of the wheels. From the position and width of the wheel pattern together with the distance measured by the laserscanner, a very accurate position and width of the object in the 3d ego-vehicles co-ordinate system can be calculated. As this wheel pattern is only exhibited by cars in the far field other featured are taken into account.

Image Edges

Figure 18c shows the horizontal, vertical and combined edge estimates of the ROI. The ROI is searched for strong support of vertical edges for the estimation of the left and right edges of the vehicle. Different hypothesis are weighted with the expected width and position of the vehicle in the image domain calculated from the laserscanner measurements.

Symmetry

The last feature taken into account is the symmetry of the image extent in the ROI. The symmetry operator is used in order to check different possible hypothesis of left and right edge pairs for their probability of representing the true vehicle edges.

Feature Tracking

In order to stabilise the estimates a feature tracking approach is applied to the vehicle in the ROI. Additionally this tracking approach delivers an estimate for the lateral velocity if scaled with the distance measured by the laserscanner.

5   Results

A major problem of the first generation of radar based ACC systems is the inaccuracy of the estimation of the lateral offset of preceding vehicles. Therefore lane changes of objects in front of the ego-vehicle are detected only when the vehicle is already completely on the target lane. Therefore in case of an object cutting into the ego-vehicles lane, the ACC system often reacts so ate, that the user is forced to take over.

The proposed fusion was tested concerning the accuracy of the estimation of the lateral offset. Figure 19 shows the distance x of a preceding car over time. The car was detected at every instance in a distance of 50 m and slightly above 100 m. The second plot shows the width of the car estimated by the laserscanner compared to the fused estimate and the third plot shows the estimated lateral offset y. No tracking was applied in this case, every parameter was derived from a single scan and image.

The estimates of width and lateral offset of the laserscanner exhibit a certain discretisation which originates from the angular resolution. The image processing algorithms are able to refine both estimates of width and lateral offset.

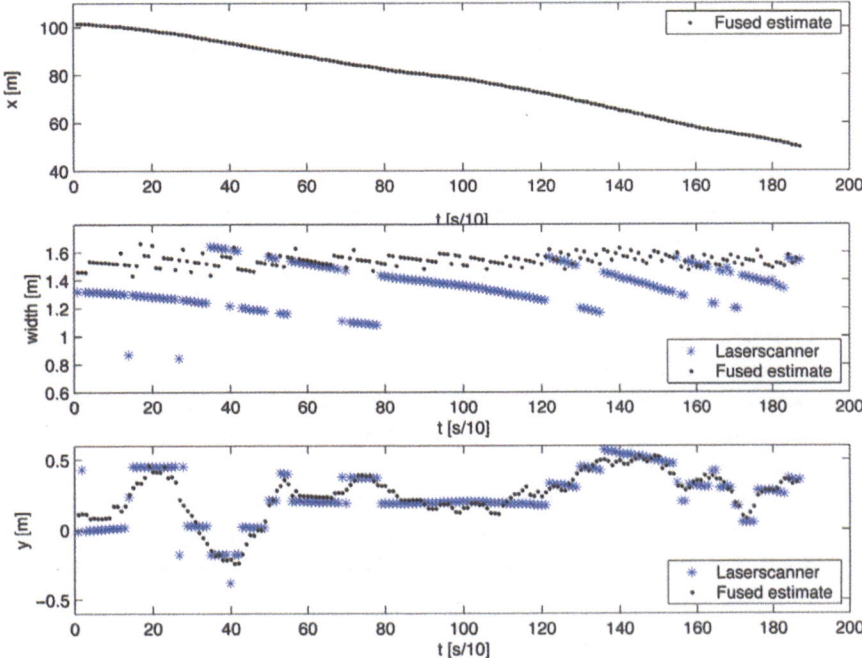

Fig. 19. Distance x, the width and the lateral offset y of a preceding car.

6   Conclusion

A sensor fusion system was presented which combines the object tracking of a laserscanner and of a video system as well as lane recognition, performed by an image processing unit. The environment recognition in the near field (<50 m) is mainly based on the very accurate laserscanners object tracking which covers an angular field of view of 180°. In the frontal far field the lateral offset estimate is refined by an image processing module. The fused estimate exhibits a high precision which is necessary for early lane change detection of preceding vehicles. The environment description is application independent and can be used by multiple automotive active safety and comfort applications.

References

[1] Kay Fuerstenberg, Pierre Baraud, Gabriella Caporaletti, Silvia Citelli, Zafrir Eitan, Ulrich Lages, and Christoph Lavergne, "Development of a Pre-Chrash sensorial system — the CHAMELEON Project," in VDI Berichte 1653: Fahrzeugkonzepte für das 2. Jahrhundert Automobiltechnik 2001, Wolfsburg, Germany, 2001, pp. 289–310.

[2] Dr. Jochen Langheim, A.J. Buchanan, Ulrich Lages, and M.Wahl, "CARSENSE— New environment sensing for advanced driver assistance systems," in Proceedings of the IEEE Intelligent Vehicle Symposium 2001, Tokyo, Japan, 2001, pp. 89–94.

[3] C. Stiller, J. Hipp, C. Rössig, and A. Ewald, "Multisensor obstacle detection and tracking," Image and Vision Computing, vol. 18, pp. 389–396, 2000.

[4] Domninique Gruyer, Cyril Royere, and Veronique Berge-Cherfaoui, "Credibilist multi-sensor fusion for the mapping of dynamic environment," in Fusion 2000, Paris, France, July 2000.

[5] A. Vukotich and Alexander Kirchner, "Sensor fusion for driver-assistance-systems," in Elektronik im Kraftfahrzeug, Baden-Baden, Germany, 2001.

[6] K.C.J. Dietmayer, A. Kirchner and N. Kaempchen, "Fusionsarchitekturen zur Umfeldwahrnehmung für zukünftige Fahrerassistenzsysteme," in Fahrerassistenzsysteme, Springer Verlag, Germany, 2003, accepted

[7] N. Kaempchen and K. C. J. Dietmayer, "Data synchronization strategies for multi-sensor fusion", in Proceedings of the 10th World Congress on Intelligent Transport Systems and Services, Madrid, Spain, 2003, No. T2250

[8] Fuerstenberg, K.Ch.; Dietmayer, K.C.J; Eisenlauer, S.; Willhoeft, V.: Multilayer Laserscanner for robust Object Tracking and Classification in Urban Traffic Scenes. Proceedings of ITS 2002, 9th World Congress on Intelligent Transport Systems, October 2002, ITS 2002 Chicago, Paper 2054.

[9] Fuerstenberg, K.Ch.; Dietmayer, K.C.J.; Lages, U.: Laserscanner Innovations for Detection of Obstacles and Road. Proceedings of AMAA 2003, 7th International Conference on Advanced Microsystems for Automotive Applications, May 2003, AMAA 2003 Berlin, Germany.

[10] Bar-Shalom, Y.; Li, X.R.: Estimation and Tracking, Principles, Techniques, and Software. Artech House, Boston, 1993.

[11] Welch, G.; Bishop, G.: An Introduction to the Kalman Filter. University of North Carolina, Department of Computer Science. http://www.cs.unc.edu/~welch

[12] Fuerstenberg, K.Ch.; Dietmayer, K.C.J.; Willhoeft, V.: Pedestrian Recognition in Urban Traffic using a vehicle based Multilayer Laserscanner. Proceedings of IV 2002, IEEE Intelligent Vehicles Symposium, IV 2002 Versailles, Paper IV-80.

Nico Kaempchen, Kay Ch. Fuerstenberg, Alexander G. Skibicki, Klaus C. J. Dietmayer
Department of Measurement, Control and Microtechnology
University of Ulm
Albert-Einstein-Allee 41
89081 Ulm
Germany
Nico.Kaempchen@e-technik.uni-ulm.de

Keywords: Preventive Safety, Driver Assistance, Sensor fusion, Laserscanner, Video

Dealing with Uncertainty in Automotive Collision Avoidance

J. Jansson, Volvo Car Corporation

Abstract

Modern automobiles incorporate more and more active driver assist system that uses sensors and microprocessors to control the vehicles dynamics. This paper discus a driver assist system that performs autonomous braking when the vehicle is close to colliding. Decision making in such systems is inherently uncertain, due to the sensors' measurement uncertainty and the uncertainty of the driver's future actions. Furthermore computational capacity is limited by the microprocessors used in automotive applications. In this paper considerations for dealing with the uncertainty of the estimated parameters in the decision making process is discussed. Risk metrics and a computationally efficient method for decision making is proposed. It will be shown that under certain conditions the risk for a to early intervention can be kept constant for different closing velocities using the proposed method. Furthermore tracking and modelling of driver, sensors and brake system is discussed. The models of driver actions and of the radar sensor measurement error are based on measurements from a Collision Avoidance system equipped Volvo V70, provided by Volvo Car Corporation.

1   Introduction

A current trend in automotive industry is to introduce active safety systems that avoid or mitigate collisions. One system with a potential large positive impact on accident statistics is forward collision avoidance systems (FCAS), using sensors such as radar, lidar and cameras to monitor the region in front of the vehicle. A tracking algorithm is used to estimate the state of the objects ahead and a decision algorithm uses the estimated states to determine any action. In figure 1a schematic picture over a FCAS is presented.

In this paper we will specifically discuss how to deal with the uncertainty of the estimated parameters in decision making process of a FCAS. Specifically a system that performs late braking to reduce the collision speed, which is referred to as collision mitigation by braking system (CMbB system), will be considered. References to this type of system can be found in [3]. There are several motivations for FCAS. One is their potential ability to affect rear-end collisions which constitute approximately 30% [6] of all collisions. Furthermore human factors contribute to approximately 90% of all traffic accidents [5]. For reasons such as driver acceptance and legal requirements of the system not to make faulty decisions, the tracking and decision algorithms are crucial.

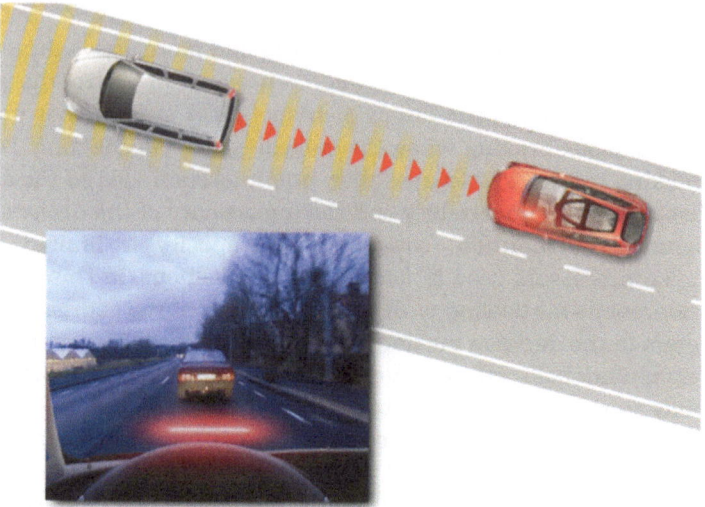

Fig. 1.   Illustration of a forward collision avoidance system. This system warns the driver of a potential collision, by means of flashing lights reflected in the windshield.

2   Tracking and Vehicle Modeling

For a CMbB system the measurement noise (radar, lidar camera etc.) and the process noise (driver inputs) are not necessarily Gaussian. The process model of a car's motion is in general non-linear. A standard approach (which will be adopted in this paper) to deal with the nonlinearities is to make time varying linear approximation of the system and apply the time varying Kalman filter (this also requires the noise distributions to be approximated with Gaussian distributions).

2.1 Tracking Model

The following state space model for the vehicle dynamics and sensor measurements will be used:

$$x_{t+1} = F_t x_t + B_{u,t} u_t + B_w w_t \tag{1a}$$
$$y_t = H_t x_t + e_t. \tag{1b}$$

For CMbB, the state vector, $x_t \in R^n$, contains relative position (longitudinal and lateral) to the other vehicle of potential collision risk, and relative velocity. Other quantities that might be estimated are relative acceleration and yaw-rate. Further, w_t denotes the driver inputs from the accelerator, brake and steering wheel. The driver is thus modeled by the probability density distribution $p_w(.)$ of the stochastic variable w_t. Since it also includes actions in the other car, w_t has to be considered as un-measurable. The measurement relation (1b) comes from distance and bearing sensors such as radar, lidar and camera. The measurement noise, e_t, includes clutter, multi-path, multiple reflection points and other artifacts introduced in the low-level signal processing. The probability density of e_t, $p_e(.)$ is used to model the sensor noise properties. If we only consider longitudinal motion we can use the following model

$$x_t = \begin{pmatrix} \tilde{p} \\ \tilde{v} \\ \tilde{a} \end{pmatrix}, u_t = \begin{pmatrix} a_{ref} \end{pmatrix} \tag{2}$$

Here tracking will be performed in a relative coordinate system, the states are relative position $\tilde{p}$, velocity $\tilde{v}$ and acceleration $\tilde{a}$. The control signal a_{ref} is the requested acceleration of the host vehicle. A more thorough discussion on how to model the brake system follows below. The tracking model with sample time T is then

$$x_{t+1} = A x_t + B_{u,t} u_t + B_w w_t \tag{3a}$$
$$y_t = h(x_t) + e_t, \tag{3b}$$

with

$$A = \begin{pmatrix} 1 & T & T^2/2 \\ 0 & 1 & T \\ 0 & 0 & e^{-k_1 T} \end{pmatrix}, B_{u,t} = \begin{pmatrix} 0 \\ 0 \\ 1 - e^{-k_1 T} \end{pmatrix}, B_w = \begin{pmatrix} T^3/6 \\ T^2/2 \\ T \end{pmatrix}, h(x_t) = \begin{pmatrix} \tilde{p} \end{pmatrix} \tag{4}$$

2.2 State Noise Model

When using the EKF for target tracking, one assumes a Gaussian distribution for w_r. The main source of the state noise (w_r) is the driver. However the input from the driver is limited by the performance of the vehicle. Factors that play a significant role are available tire-to-road friction, brake- and steering-system characteristics and engine power. A vehicle's dynamical properties change with time and the driver's behavior also typically varies with road/traffic condition. We try to model all these phenomena with a Gaussian noise, clearly a very coarse model. To get realistic values on the process noise, data was collected during normal driving in rural traffic. A histogram of accelerations during this drive is displayed in figure 2.

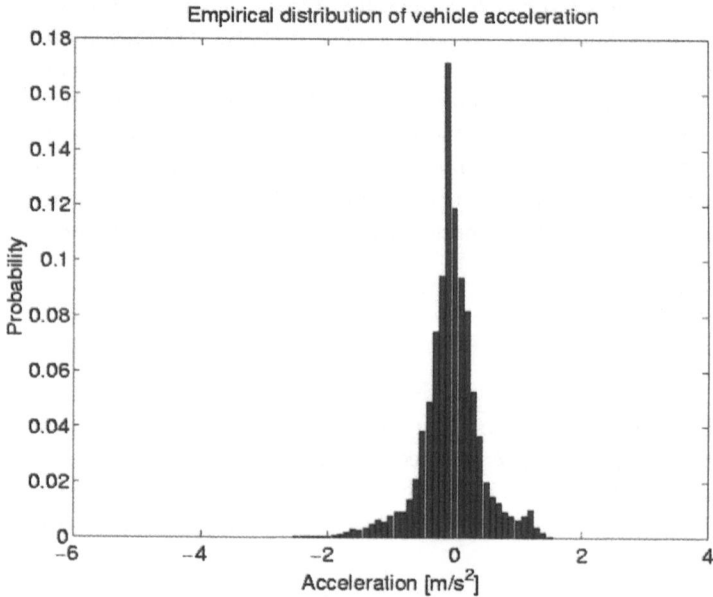

Fig. 2.   Probability density $(p_w(a_{host}))$ for acceleration of a vehicle driven in urban traffic for 45 minutes.

2.3 Measurement Noise Model

The measurement equation involves range and azimuth angle measurements to the vehicle ahead. Similarly to the above, the range error distribution can be modeled. A common model is to assume Gaussian noise. Figure 3 shows the histogram of radar range measurements.

The data collection was performed with the target vehicle (a Volvo V70) towing the radar equipped vehicle at constant speed.

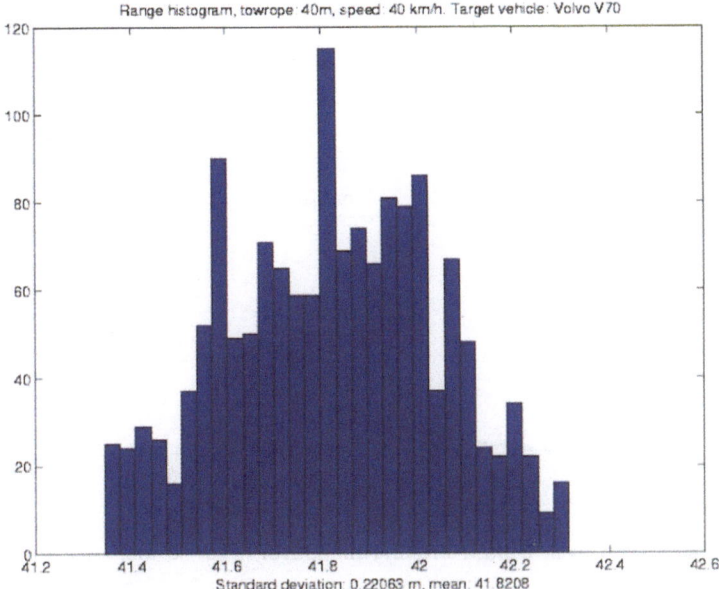

Fig. 3.   Probability density of radar measurement $p_e(\tilde{p})$ at constant speed. The measurement where performed with the target vehicle towing the radar host vehicle. To minimize range variation due to rope elasticity the test where performed on a test track with even pavement and cruise control to keep the speed constant.

2.4   Brake System Model

To calculate the collision speed the brake system characteristics need to be known. A complete brake system model is very complex. However looking at overall vehicle performance a simple yet accurate description can be obtained using a first order system for the acceleration. Figure 4 shows measurement data from 10 brake maneuvers with a Volvo V70 and a first order approximation (according to (5)).

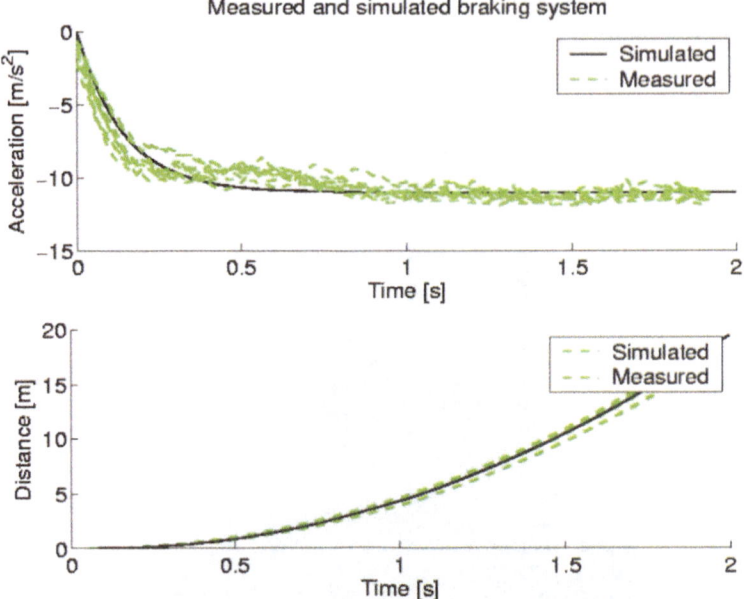

Fig. 4. Data from a hard braking maneuver, 110 - 0 km/h with a Volvo V70.

The measurements were performed on warm and dry asphalt. For the simulated data a first order system (5) was used, $k_1=7$ and the input $a_{ref}=11$ m/s^2 was chosen such that the system rise time is 0.3 s and the stationary value is 11 m/s^2.

$$A(s) = G_{brake}(s)A_{ref}(s) = \frac{k_1}{s + k_1}A_{ref}(s) \tag{5}$$

Comparing traveled distance after 2 seconds we find that the difference is less than 1 m which is sufficient for our purposes. More information on typical brake system behavior can be found in [2].

3 Statistical Decision Making

Statistical decision making requires an estimate of the distribution of x_t from measurements y_t. A general model-based statistical decision criterion for the risk metric $g(x)$ takes the form

$$Pr\{g(x_t) > g_{th}\} > 1 - \alpha \tag{6}$$

where α determines the confidence level. Note that this is a more flexible approach than a rule like

$$g(x_t) > g_{th} \tag{7}$$

since in (6) we use the complete a posteriori distribution of the state vector and not only the estimate $\hat{x}_t$. How successful is the rule (6) in mitigating a collision? That depends of course on the rule $g(x)$ itself, but also on how accurately the probability density function (PDF), $p(x)$, reflects the true distribution of the relative distance between the vehicles. Which rule should be used (6) or (7)? Both approaches have their pros and cons. While (6) has the advantage of a fixed confidence level, for making a decision, an analytical solution is in general hard to find. Thus a computationally expensive algorithm is often required to give a numerical approximation [4]. The rule (7) on the other hand is computationally cheap, but if $p(x)$ varies a lot the probability for making the decision might also vary significantly. In this paper we will study a third alternative decision criterion. The aim is to keep computational complexity low while taking into account the time varying uncertainty of the decision quantity. We will use the following criterion.

$$g(\hat{x}_t) - C_1 B(g(x_t)) > g_{th} + C_2 D(g(x_t)) \tag{8}$$

Here $B(.)$ and $D(.)$ denote bias and standard deviation respectively and C_i are design parameters. One motivation for (8) is the well known statistical properties of the Gaussian distribution. If $g(\hat{x}_t)$ is Gaussian, (8) corresponds to making the decision with a certain confidence i.e. the same as in (6).

3.1 Estimating Statistics

How is $B(g(\hat{x}_t))$ and $D(g(\hat{x}_t))$ calculated? The bias and standard deviation of (8) is defined in the following way

Bias : $B(g(x)) = E(g(x)) - g(x)$ (9)

Standard Deviation : $D(g(\hat{x}_t)) = \sqrt{E(g^2(x_t)) - E(g(x_t))^2}$ (10)

where $E(.)$ denotes the expectation value. $E(g^j(\hat{x}, t))$ is known as the jth order zero point moment of $g(x_t)$, which is defined by

$$E(g^j(x_t)) = \int_{R_n} g^j(x_t)p(g(x_t))dx_t$$ (11)

$p(g(x_t))$ is the PDF of $g(x_t)$. Considering that $g(.)$ can be a quite complicated function the integral (11) or even the PDF $p(g(x_t))$ might be hard to compute analytically and computationally costly to approximate by numerical integration. Hence we will use an approximation of the expectation value in (11), where an assumption of a Gaussian distributed x_t is used. The approximation is obtained from the expectation value of a Taylor expansion of $g^j(x_t)$ given by

$$g^j(x_t) = g^j(x_t) + \sum_{i=1}^{n} \frac{\partial g^j}{\partial x_i}\bar{x}_i + \sum_{i=1}^{n}\sum_{j=1}^{n} \frac{\partial^2 g^j}{\partial x_i \partial x_j}\bar{x}_i\bar{x}_j + \dots$$ (12)

where $\bar{x}_i = x_i - \hat{x}_i$ and x_i denotes the ith element of the state vector x_t. An approximation of the expectation $E(g^j(x_t))$ can thus be found from the expression

$$E(g^j(x)) = g^j(\hat{x}) + \sum_{i=1}^{n} \frac{\partial g^j}{\partial x_i}E(\bar{x}_i) + \sum_{i=1}^{n}\sum_{k=1}^{n} \frac{\partial^2 g^j}{\partial x_i \partial x_k}E(\bar{x}_i\bar{x}_k) + \dots$$ (13)

By choosing how many terms of the series expansion to consider the accuracy and computational burden can be traded off. If we neglect terms higher than the second order we of course get the well known Gauss approximation formula. To evaluate (13) it is clear that higher order moments of x_i need to be known. In the case these can be calculated from the characteristic function

$$\Psi(t) = e^{b^T \hat{x} + \frac{x^T P x}{2}}$$

Here $b=[b\_1\ b\_2\ ...b\_n]^T$ and P is the covariance matrix of x. The jth order moments of x_i are given by

$$E(x_i^j) = \frac{\partial^j \Psi(t)}{\partial b_i^j}\Big|_{b=0} \tag{15}$$

Thus inserting the approximation from (13) into (9) and (10) an approximation of $B(g(x_t))$ and $D(g(x_t))$ can be calculated. The same reasoning can of course also be used to estimate other interesting statistical properties such as e.g. skewness and kurtosis.

3.2 Application to Collision Avoidance

In this paper we will study a CMbB system, which performs maximum braking when a collision is becoming imminent. By imminent we mean that a maneuver that is close to or exceeds the vehicles handling limits is required to avoid collision. To simplify the analysis only longitudinal motion will be considered, in general both lateral and longitudinal motions need to be considered. A vehicle's handling properties are often measured in terms of acceleration. One appropriate risk metric for a CMbB system is to calculate the acceleration required to avoid collision. Under the assumption that the tracked vehicle continues with constant acceleration the relative required acceleration is given by the solution to

$$0 = \tilde{p}_t + \tilde{v}_t t_0 + \frac{\tilde{a}_t t_0^2}{2} \tag{16}$$

where t_0 is the time to accelerate to zero relative velocity and $\tilde{p}$, $\tilde{v}$, $\tilde{a}$ are relative position, velocity and acceleration respectively

$$t_0 = -\frac{\tilde{v}_t}{\tilde{a}_t} \tag{17}$$

The host required longitudinal acceleration will thus be given by

$$a_{h,t} = a_{o,t} - \tilde{a}_t = a_{o,t} - \frac{\tilde{v}_t^2}{2\tilde{p}_t} \tag{18}$$

Here $a_{h,t}$ and $a_{o,t}$ denotes host and tracked vehicle acceleration. The equation (18) was derived assuming that the maximum deceleration can be achieved momentaneously. Looking at figure 4 it is clear that this is not entirely true. A better risk metric also considers the brake system rise time. Using the model (5) for the brake system we get the following expression for the relative acceleration (again we assume constant acceleration for the tracked vehicle)

$$\tilde{a}(t) = a_{ref} - a_{ref}e^{-k_1 t} \tag{19}$$

where a_{ref} is the input requested acceleration. The velocity and traveled distance during deceleration is given by

$$\tilde{v}(t) = \tilde{v}_0 - \frac{a_{ref}}{k_1} + a_{ref}t + \frac{a_{ref}e^{-k_1 t}}{k_1} \tag{20}$$

$$\tilde{p}(t) = \frac{a_{ref}}{k_1^2} + (\tilde{v}_0 - \frac{a_{ref}}{k_1})t + \frac{a_{ref}t^2}{2} - \frac{a_{ref}e^{-k_1 t}}{k_1^2} \tag{21}$$

The time to bring the velocity to zero is given by the solution to (19) which yields

$$t_{stopp} = -\frac{(\tilde{v}_0 k_1 - a_{ref} - L(-e^{\frac{(\tilde{v}_0 k_1 - a_{ref})}{a_{ref}}})a_{ref})}{a_{ref}k_1} \tag{22}$$

where $L(x)$ denotes the solution to $we^w = x$. Combining (22) and (21) allows the stopping distance to be calculated. The new risk metric will thus be the stopping distance calculated for $a_{ref} = a_{min}$, this is compared to the measured distance, to determine any action. We shall now investigate performance of a risk metric based on maximum deceleration when used as a decision criterion according to (7) and (8). To keep the analysis simple the brake system rise time will be neglected, and thus (18) will used be as risk metric.

4   Simulations

To evaluate the different decision criteria, a scenario where a CMbB system equipped vehicle is approaching an object at constant velocity according to figure 1, will be studied. This type of scenario corresponds to more than 30% of all rear-end collisions [1]. First we will look at the probability of performing an intervention during the sample interval when the CMbB vehicle is exactly at the intervention boundary i.e. at the instant when a collision is becoming imminent. Here we define a collision as being imminent when a deceleration of 8 m/s² or more is required to avoid colliding. In figure 5 the probability of an intervention for the criterions (7) and (8) can be compared at different closing velocities, the parameters for the calculation can be found in table 1.

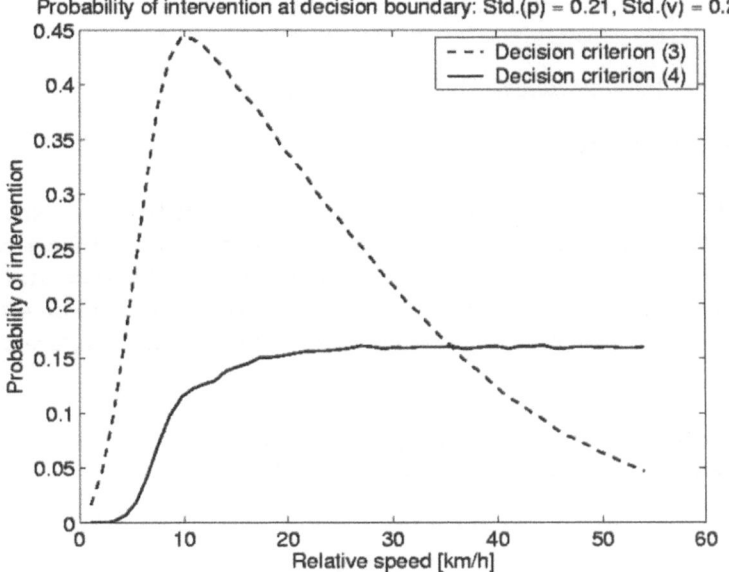

Fig. 5.   Probability of performing a CMbB intervention at the sampling interval when the collision imminent boundary is crossed. The required acceleration is -8 [m/s²]

In figure 5 the probability of an intervention is displayed only for a certain time instant (i.e. when the collision is becoming imminent). Next we will study system performance in terms of collision speed and probability of making a too early intervention for the head-on scenario. A too early intervention is defined as a CMbB system induced brake maneuver before the collision imminent boundary is reached. The process noise, *w*, and measurement noise, *e*, are assumed to be independent and Gaussian. The values of the noise variances were chosen to correspond to the measured pdfs of section 2. The brake system model is according to the model described in section 2.4, however the maximum deceleration is somewhat lower, since the perfect conditions on the test tracks are not always present on a normal road. Both decision criterion functions are applied to the tracking system output. A summary of the simulation parameters is given in table 1.

SIMULATION SUMMARY:
Decision criterion (7) parameters: $g_{th} = -8.5$ [m/s^2].
Decision criterion (8) parameters: $C_1=1$, $C_2=1$ $g_{th} = -8$ [m/s^2].
Sensor and tracking system update rate: 10 Hz.
Measurement noise: $e_t \in N(0, 0.25)$
Brake system model: $k_1 = 7$ (rise time 300 milliseconds).

Table 1. Summary of important simulation parameters.

In figure 6 the simulation results are displayed for speeds from 5-60 km/h. For each speed 2000 Monte Carlo simulations were performed, and for each scenario both decision criteria were applied to the same tracking data. At higher speeds than 60 km/h a steering avoidance maneuver is much more efficient than a braking maneuver, therefore higher velocities are not as interesting to examine.

Figure 7 displays the same simulations scenarios as figure 6, but here the standard deviation of the measurement noise was doubled.

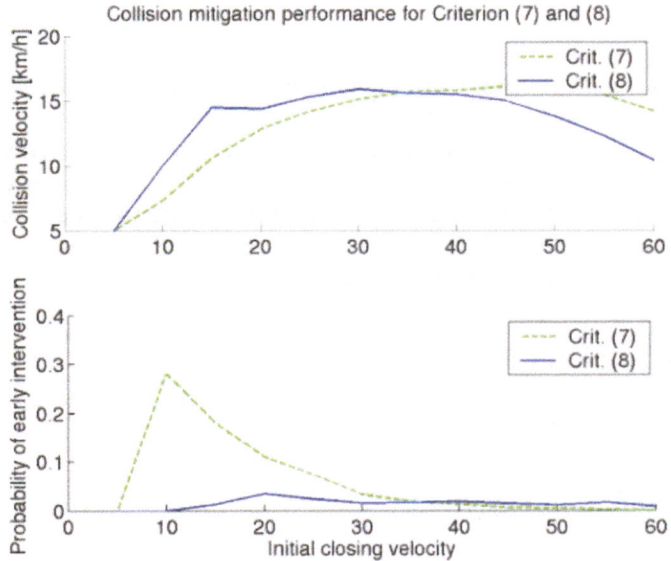

Fig. 6.   Head-on scenario simulation results for Criterion (7) and (8) using the risk metric (18). The simulation parameters are summarized in Table 1

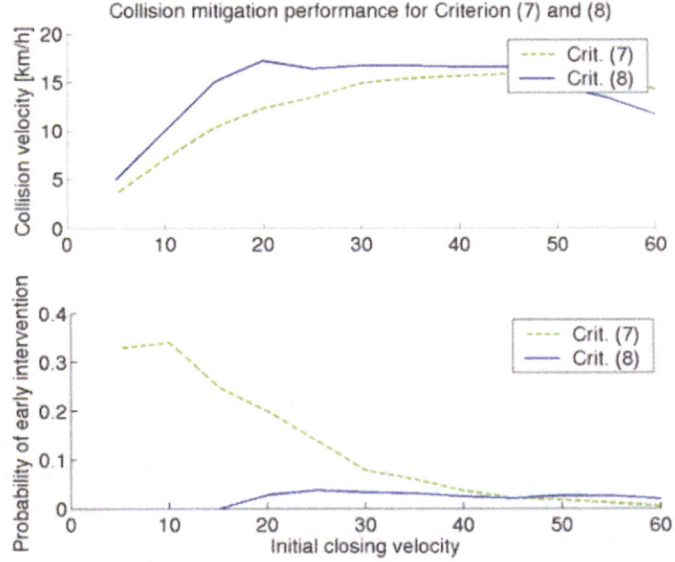

Fig. 7.   Head-on scenario simulation results for Criterion (7) and (8) using the risk metric (18). Here the std. of the measurement noise was doubled (compared to figure 6).

5 Analysis

How well does the criterion (7) work compared to (8)? In figure 5 the probability of an intervention at a certain time instant is compared for different velocities. From the figure it is clear that for the criterion (7) the probability for an intervention vary while for (8) it is kept constant (for speeds above 15 km/h). Why is it that the decision criterion (8) manage to keep a constant risk level? To understand this we need to look at the criterion more closely. We rewrite the estimated required acceleration as

$$\hat{a}_{host} = \hat{a}_{obj} - \hat{a} = a_{\hat{obj}} - \frac{\hat{v}^2}{2\hat{p}}. \tag{23}$$

$$= a_{obj} + w_a - \frac{(\tilde{v} + w_v)^2}{2(\tilde{p} + w_p)} \tag{24}$$

$$= a_{obj} + w_a - \frac{(\tilde{v}^2 + 2\tilde{v}w_v + w_v^2)}{2(\tilde{p} + w_p)} \tag{25}$$

where in our case w_p, w_v and $w_a \in N(0,\sigma_i)$. When $\tilde{p}$ and $\tilde{v}$ are large compared to the noise terms w_v and w_p (18) is well approximated by

$$\approx a_{obj} + w_a - \frac{(\tilde{v}^2 + 2\tilde{v}w_v)}{2\tilde{p}} + \frac{\tilde{v}^2}{2\tilde{p}^2}w_p + \frac{\tilde{v}^2}{4\tilde{p}^2} \tag{26}$$

Equation (26) has a Gaussian distribution since it consists of sum of Gaussian. Thus, when $\tilde{p}$ and $\tilde{v}$ are large compared to the uncertainty of the risk metric (18) we approximately have

$$\hat{a}_{host} \in N(a_{obj} + \frac{\tilde{v}^2}{2\tilde{p}} + \frac{\tilde{v}^2}{4\tilde{p}^2}, \sigma_a + \frac{\tilde{v}}{2\tilde{p}}\sigma_v + \frac{\tilde{v}^2}{2\tilde{p}^2}\sigma_p) \tag{27}$$

The criterion (8) will thus be equivalent to the hypothesis test (6), the parameter C_2 will determine the confidence level of the test. What are the implications of this in terms of collision mitigation performance? Looking at the simulation results of figure 6, we see that for speeds of 30 km/h and above the probability of a faulty intervention is kept constant for the criterion (8). For criterion (7) on the other hand the probability of a faulty intervention decreases as the closing velocity becomes higher. At velocities around 35 km/h the performance is the same for both criteria. This is also the velocity at which they cross in figure 5. Both criterions performed bad at low speed. This is mainly caused by the fact that the dynamics of the brake system is not captured in the risk metric (18). The assumption of a constant measurement noise causes the assumption of $\bar{p}$ and $\bar{v}$ being large compared to the noise to be violated at low closing speeds. The required acceleration (18) will thus not be Gaussian. This causes the probability of intervention to vary significantly for low speeds.

6 Conclusions

In this paper statistical decision making applied to automotive collision avoidance has been discussed. A criterion that explicitly uses an approximation of the statistical properties of the risk metric has been simulated and analyzed. To compute the approximation the estimated tracking noise covariance (which was assumed to be Gaussian) was used. The main reason for this criterion was to be able to deal with changing uncertainty of the risk metric without having to evaluate the entire PDF. The criterion is motivated by properties of the Gaussian distribution, for Gaussian distributed variables the criterion corresponds to a hypothesis test. It was shown that one particular risk metric (18) can be considered to be Gaussian under certain circumstances. In head-on collision scenarios where the assumption of gaussianity holds ($v > 25$ km/h), the criterion (8) applied to (18) keeps the risk of a faulty intervention at a relatively constant level. However, a careful comparison of the results in figure 6 and 7 shows a slight change in that risk level. For the criterion (7) the risk of a faulty intervention is changed more significantly as the measurement noise is increased. For scenarios where the assumptions do not hold (e.g. at low speeds) the criterion (8) becomes too cautious, causing braking to be too late. Bad performance at low speeds are also caused by the fact that the risk metric used does not account for the brake system dynamics, here the second risk metric discussed in Section 3 can offer significant improvements.

References

[1] Automotive collision avoidance system field operational test. Technical Report DOT HS 809453, NHTSA, May 2002.

[2] D.-I. U. Adler. Automotive Handbook. Number ISBN: 0-7680-0669-4. Robert Bosch GmbH, 5th edition, 2000.

[3] J. Jansson. Tracking and Decision Making for Automotive Collision Avoidance. Licentiate Thesis no. 965, Department of Electrical Engineering, Linköping University, Sweden,September 2002.

[4] J. Jansson, R. Karlsson, and F. Gustafsson. Model-based statistical tracking and decision making for collision avoidance application. Tech. Report LiTH-ISY-R-2492, Dept of Elec. Eng. Linköping University, S-581 83 Linköping, Sweden, 2003. In English.

[5] J. Treat, N. McDonald, D. Shinar, R. Hume, R. Mayer, R. Stansifer, and N. Castellan. Tri-level study of the causes of traffic accidents. Technical report, NHTSA, 1977.

[6] B. Zhu. Potential effects on accidents from forward collision warning/avoidance system. Master Thesis LITH-ITN-EX-150-SE, ITN, Linköping University, S-581 83 Linköping, Sweden, 2001.

Jonas Jansson
Volvo Car Corporation, PVH 34
405 31 Göteborg
Sweden
jjansso1@volvocars.com

Keywords: Collision Avoidance, Collision Mitigation, Target Tracking, Statistical Decision Making, Autonomous Braking.

Paroto Project: Infrared and Radar Data Fusion for Obstacle Avoidance

Y. Le Guilloux, J. Lonnoy, R. Moreira, SAGEM SA
C. Blanc, J. Gallice, L. Trassoudaine, LASMEA
H. Tattegrain, M.-P. Bruyas, A. Chapon, INRETS

Abstract

PAROTO is a joint project investigating obstacle detection and avoidance by fusion of information from radar and infrared imagery, with the support of the French ministry of Research. The PAROTO consortium includes SAGEM, LASMEA, INRETS and TEAM.

Detection of obstacles by radar and infrared exhibit different properties so their concurrent use allows optimal performance. Innovative image processing algorithms have been set up to detect vehicles and pedestrians (see [6]). Original 77 GHz radar was also specified and realised. Fusion of information exploits a sound mathematical framework getting the best of both alarm sources. Then a filtering step eliminates pointless alarms, when it is clear that the driver has already taken appropriate actions regarding the obstacle. A dedicated driver's interface was also designed and realised to warn – visually and/or acoustically – without unnecessarily driving the driver's gaze out of the road ahead.

Both detection schemes already operate in real time. By the end of 2003, the fusion of information, integrated aboard our tested vehicle, will be evaluated together with the driver interface, providing valuable insight into automated perception dedicated to automotive safety. Results strongly incite us to anticipate the design of a low cost hardware dedicated to these functions.

1   Introduction

The aim of PAROTO system is to compensate the lack of attention of the driver, a proven source of many of accidents.

Such a system depends on several mechanisms and assumptions:

▶ A safety system should not rely on one sensor only. We have chosen two sensors: Radar and Infrared camera. They are sensitive to different physical properties of an object (materials and surface in one hand, thermal behaviour on the other). They also give complementary geometrical measures.

▶ It must be discrete but reliable (through a dedicated HMI with graduated warnings, depending on the situation).

▶ It must be accepted by the driver to be efficient.

▶ It must take into account the reaction of the driver facing the situation.

In this article, we present each functionality of the PAROTO system. We first recall the functional architecture. We describe the Radar and the Infrared camera and their related processes dedicated to obstacles detection. Then, we focus on data fusion, which builds a precise map of obstacles ahead of the vehicle. We present the driver's interface and the filtering module. We conclude with a description of the system platform.

Fig. 1.   VELAC, the vehicle of LASMEA, equipped with IR camera (red) and radar (green).

2 Functional Architecture

Both sensors feed respective processing capabilities, which in turn interact with a fusion module, the output of which is filtered in order to send only relevant signals to the driver's interface (figure 2).

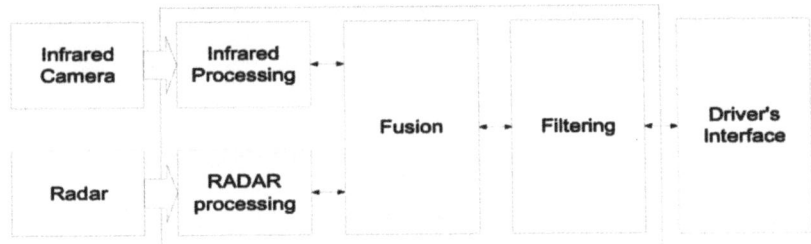

Fig. 2. Functional architecture.

3 Radar Obstacle Detection and Tracking

3.1 Radar Specifications

The key interests to use a radar in this project are on the one hand the accuracy of obstacles estimated speed and on the other hand the quality of radar data up to 150 m in spite of bad weather conditions.

According to the regulation in the European Union and in the United States, the wave emission frequency for a car anti-collision project is fixed to $f_0 = 77$ GHz (millimetre wave). The principle of the radar used in this framework is a pulsed doppler radar with a Doppler priority. Indeed, the priority is given to the speed of obstacles with respect to the experimental vehicle. The distance of the obstacles is sorted by P range gates.

Moreover, this radar has only one antenna for the emission and the reception which provide a 3.3° field of view. This system requires a circulator (ultra-high-frequency system) and a synchronisation process in order to transfer the energy either between the transmitter and the antenna or between the antenna and the receiver (figure 3). The reception channel being very sensitive, the system must perfectly isolate this channel from the emission channel because of the strong emitted power.

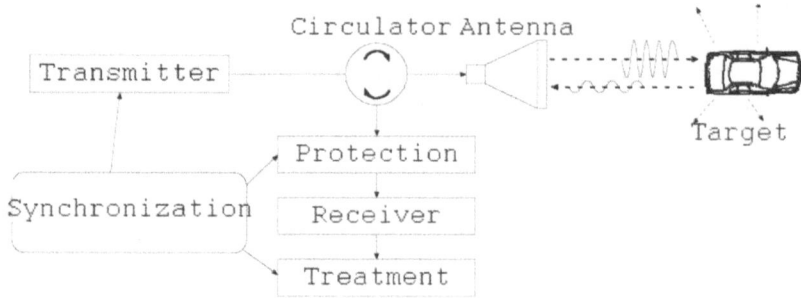

Fig. 3.   Structure of the PAROTO radar with one antenna.

The radar technique used in this project consists in sending by repetitive pulses (width t_e = 150 ns and period T_r = 4 µs) a sinusoidal electromagnetic wave and to receive and analyse the signal reflected by a target (see figure 4). The data resolution is:

$$\partial D = \frac{ct_e}{2} = \frac{3 \cdot 10^8 \cdot 150 \cdot 10^{-9}}{2} = 22.5\text{m} \tag{1}$$

$$\partial v = \frac{c}{2 \cdot f_0 \cdot T_r \cdot N} = \frac{3 \cdot 10^8}{2 \cdot 77 \cdot 10^9 \cdot 4 \cdot 10^{-6} \cdot 2048} = 0.238\text{m/s} = 0.856\text{km/h} \tag{2}$$

where N represents the period number (actually 2048 which correspond to 8 ms of time analysis) of the pulse repetition T_r.

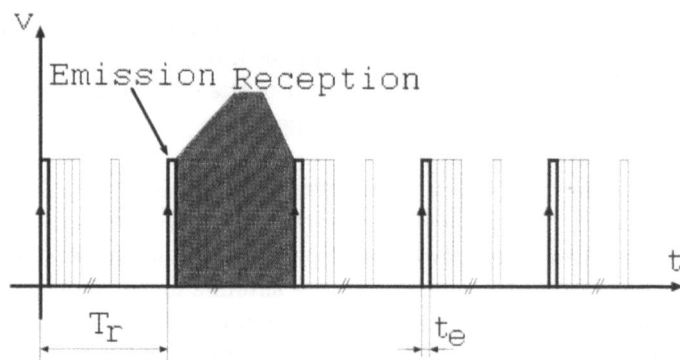

Fig. 4.   Emission and reception chart.

3.2 Obstacle Detection and Tracking

Echo extraction [1]

The process first acquires, each $n \times T_r(n \in [0,....N-1])$ time, an array of $P \times N$ (3) data corresponding to the sample of the two I, Q reflected signal (0°, 90°) provided by a high speed 40 MHz two lines converter stage. Then we compute an estimate of the power spectral density by FFT (Fast Fourier Transform) analysis.

In this new array, we apply a threshold-clustering algorithm and extract echoes.

An echo is then defined by four parameters:
► time:

$$n \times T_r \left(n \in \left[0,...,N-1 \right] \right) \tag{4}$$

► amplitude: value of the power spectral density versus frequency
► Doppler: position in the column's power spectral density array

$$V_r = \left(colunm - \frac{N}{2} \right) \cdot \partial v \tag{5}$$

► range: position in the line's power spectral density array

$$r = \left(line - 1 \right) \cdot \partial D + \frac{\partial D}{2} \tag{6}$$

Kalman filter tracking [2]

A constant velocity model is used into a Kalman filter to track echoes. This tracking is performed into three steps:
► Initialisation of tracks (r, V_r) using coherence in many frames,
► filtering of tracks: prediction and estimation,
► data association.

This tracking yields a range estimate more accurate than the gate value (22.5 m), and also obstacle speed, as it is shown on following results.

3.3 Some Results

On the following illustrations we present some results of our tracking in foggy conditions.

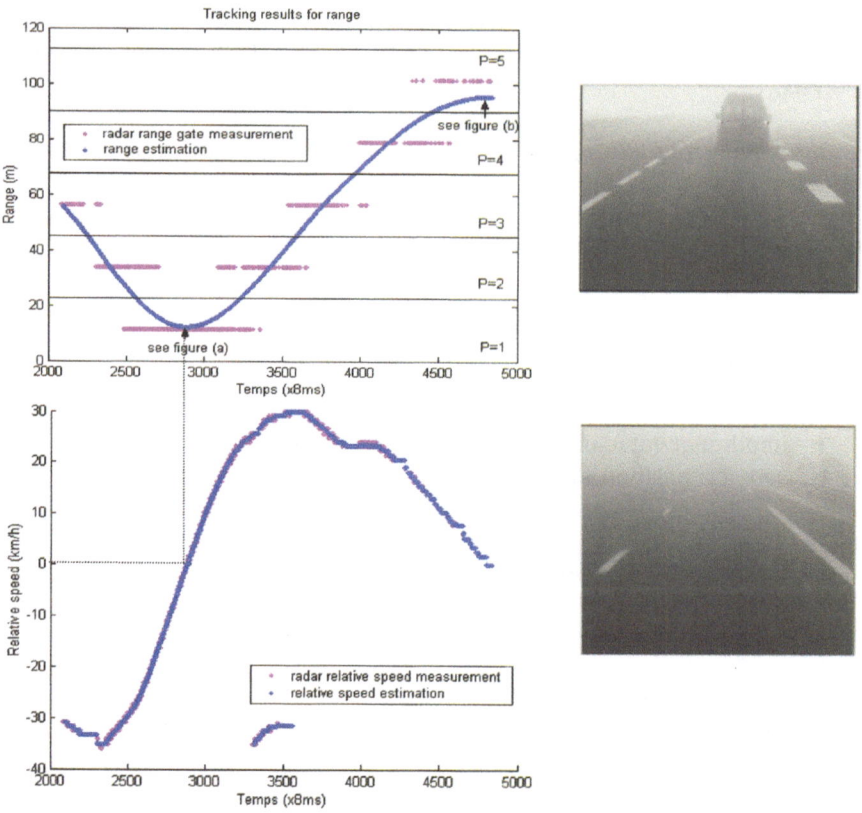

Fig. 5. Tracking results.

On these illustrations, we show that the tracking algorithm provides a range estimate more accurately than a range gate. For the relative speed, we can see that the estimate corresponds to radar measurement. Furthermore, we can see that when obstacle's relative speed becomes positive the range increases. Finally as we can see, we are able to track an obstacle even if it is not visible (figure 5b).

3.4 Implementation

In order to validate this approach in a real navigation context, this Radar has been implemented on the experimental vehicle (VELAC) of the LASMEA Laboratory (figure 6).

Fig. 6. Radar location on VELAC.

The acquisition of the analogical data from the Radar is carried out on a standard PC equipped with an acquisition card AD-LINK PCIS-9812.

4 Infrared Obstacle Detection and Tracking

We first present the sensor technology and its location on the vehicle. Then we recall the infrared image processing. We conclude this section with some new results.

4.1 Infrared Camera Specifications

Some car manufacturers already propose, or at least evaluate infrared cameras.

Our infrared sensor is an uncooled 8-12 micro bolometer. Its thermal resolution is 80 mK. The size of images is 320 x 240 pixels. The optical design allows a 40° x 30° field of view.

The camera is located on top of the windscreen, for practical and theoretical reasons (dirt, projections and geometry of distance estimation).

4.2 Infrared Image Processing

What's hot:
In classical road situations, the sources of danger relate to human activity. This leads us to focus the detection on vehicles, cycles and pedestrians, which exhibit strong and typical behaviours in infrared.

Fig. 7.   IR aspects of road users.

Detection and tracking:
Detection of objects of interest is done in two steps:
  ▶ Extraction of various cues from image, using low-level processing modules. One of these modules extracts significant hot regions, which are typical products of infrared image processing. Other cues are based on geometrical criteria, gradients shapes, symmetries,
  ▶ Merging of low-level cues into objects, assuming loose geometrical templates of those objects.

Then positions of detected objects are mapped onto the road, under the assumption that the road belongs to a plane. Since this operation is sensitive to the changes of pitch, motion analysis – based on feature tracking – is performed in order to estimate the pitch and correct the 3D positions accordingly.

At this point, we track objects using a constant velocity Kalman filter.

Some results: Vehicle detection
On the following illustrations, black boxes are frame-by-frame detected vehicles since tracking can only be shown on live demonstration. The algorithm works in real time. We have experimented it in various situations, as suggested below.

Fig. 8.   Vehicle detection in a rural context.

Fig. 9.   Vehicle detection on highway.

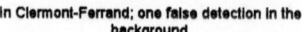

In Clermont-Ferrand; one false detection in the background   Stopping at traffic lights; on the right hand side, two vehicles are merged into one detection

Initiating a left turn

Fig. 10. Vehicle detection in an urban context.

Figure 10 shows results in an urban context. Complex situation implies many possible sources of false detection.

Pedestrian Detection in Various Contexts

The pedestrian detection process has not reached such a completion status as the vehicle detection one. However, evaluated on many sequences at hand, our low-level detection modules have performed quite well, as the following images illustrate it.

Note that at this point, a same pedestrian may be detected in two parts, due to the still crude merging of our low-level cues.

Pedestrians in an simple context   Pedestrian walking on our lane

Urban context   Rural context, pedestrian crossing our lane

Fig. 11.   Pedestrian detection.

5   Data Fusion

Whereas radar is able to tell at what distance it points echoes, the direction being somewhat fuzzy, Infrared can give the direction in which an event is detected, but no precise access to the distance at which it occurs. Information coming from those sensors complements each other.

Data from both sensors are first processed separately. Processes go as far as possible in the interpretation of the scene, to extract the most relevant information from sensors. Then the outputs of processes – the tracks of detected objects – are passed to the fusion module.

From those tracks, the fusion module builds a map describing the situation ahead of the vehicle. Once built, this map is sent to the filtering module, the goal of which is to determine whether the driver should be alerted or not.

5.1   Areas of Fusion to Consider

Radar and infrared cover complementary areas ahead of the vehicle (figure 12). We call a track or an area "dual mode" when both sensors are involved.

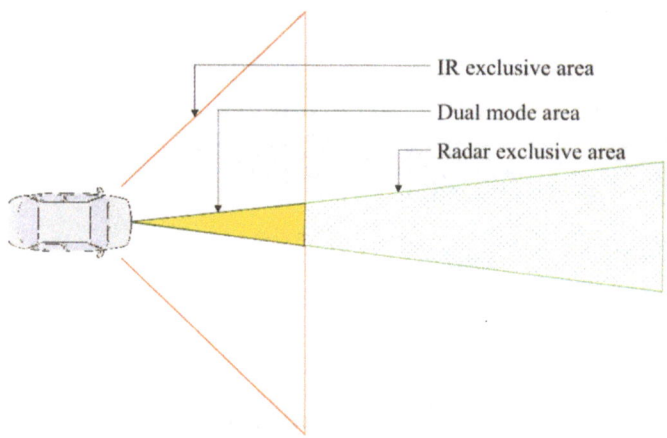

Fig. 12.   Areas covered by PAROTO system.

Tracks produced by single processes lie in one of the three areas described in figure 12. If an IR (resp. radar) track is produced in the IR (radar) exclusive area, the fusion module cannot confirm it with a track from the other sensor process: it must rely on this single sensor. Whereas, if the track is produced in the dual mode area, a real data fusion can be performed.

5.2   Multiple Sensors, Multiple Target Tracking: a Hybrid Model of Fusion

There are three main steps in fusion process. First, we must determine which tracks can be associated. For each couple of tracks, we improve the information concerning the object represented. Finally, we maintain tracks over time.

Data Association

Observations from IR and radar are dated with respect to a common clock, so they can be compared. When trying to associate an IR track with a radar track in the dual mode area, tracks are extrapolated at the current time and compared.

The comparison is done on relative distance and speed, which compose the common space of measurements.

We assume that measurements are given with a Gaussian error. Then, we calculate a Mahalanobis distance between the IR track and the radar track and compare the result with a threshold to decide if the tracks refer to the same object.

Data Consolidation

A couple of IR and radar tracks matched by the association step are combined into a new dual mode track. A Kalman filter computes the position and velocity of this dual mode track, which are given with better accuracy than those of its IR and Radar components.

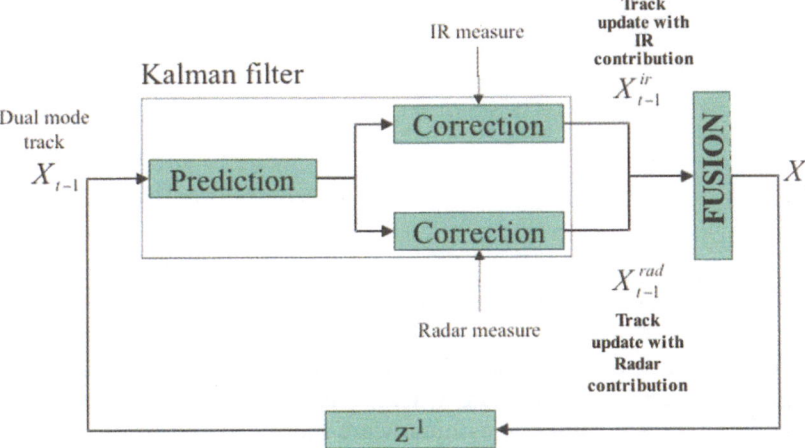

Fig. 13.   Data consolidation, improvement of measures.

Dual Mode Track Management

A dual mode track can disappear in the following situations:

▶ a sensor track (or both) disappear,

▶ associated tracks do not represent the same object and diverge from each other. To avoid keeping such a dual mode track, we regularly check the fit of the sensor tracks. If it becomes too large, we delete the dual mode track.

When a single sensor track popping in the dual mode area is not matched with any track from the other sensor, it is considered as a false alarm thus avoiding spurious detection from one sensor.

5.3 Outputs of the Data Fusion

We illustrate here some situations and show that fusion gives a significant gain in respect with separate sensor processing.

Simple Situation

Figure 14, we observe two vehicles in the image. After sensor processing step, IR and radar send to the fusion module the produced tracks.

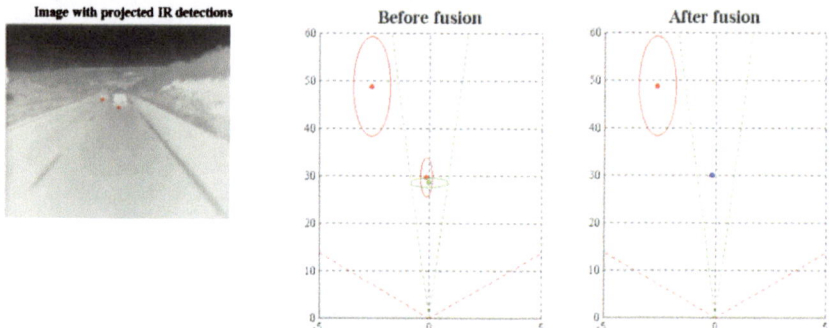

Fig. 14. Data fusion on a simple situation.

The errors on measures, symbolised by ellipsis, are as expected: IR tracks are less accurate in distance than radar tracks, and errors on IR measures grow with distance, whereas direction estimates are more accurate in IR.

The vehicle positions propose different cases:
▶ the vehicle on the left is in the IR exclusive area; its IR track cannot be confirmed by any radar track and is passed directly through fusion,
▶ the vehicle on our lane is in the dual mode area: its IR and radar tracks are close enough and associated to build a dual mode track (blue in the right hand side map).

Our observation confirms expectations: the error area of the dual mode track is far smaller than its single sensor counterparts. This is a mathematical consequence of the combination of two Gaussian distributions.

Overtaking of a Vehicle

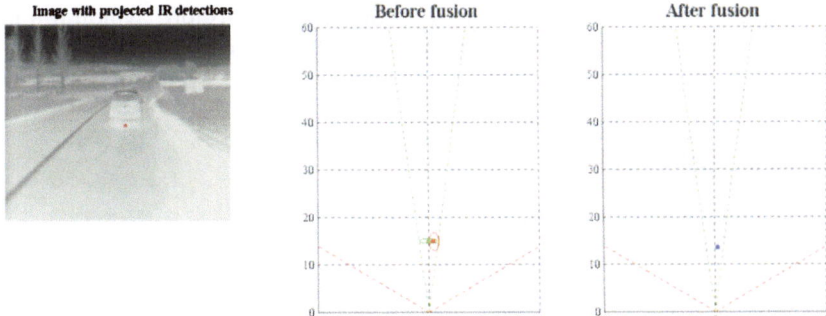

Fig. 15. Overtaking of a vehicle, step 1.

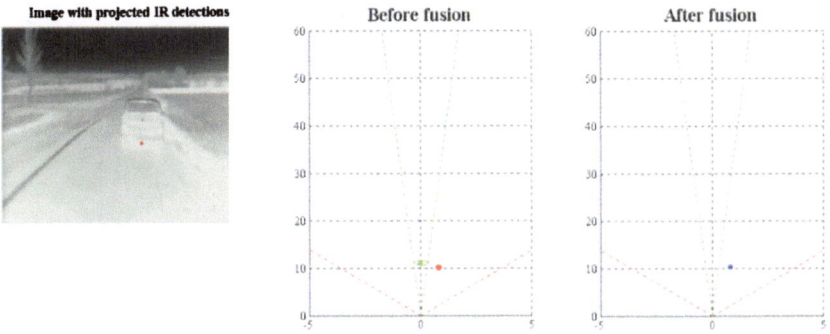

Fig. 16. Overtaking of a vehicle, step 2.

This case of overtaking illustrates the complementarity of infrared and radar. We approach the preceding vehicle (figure 15) and pull out (figure 16, at the beginning of the manoeuvre). With the radar alone, the system detects a vehicle ahead approaching until the last moment. Detection direction from infrared influences the dual mode track, allowing to detect the drift of the vehicle, therefore not classified as a threat.

Note: although the vehicle seems to be detected out of the radar field in figure 16, is due to the point wise representation, the radar actually detects the left side of the vehicle which is overlapping the dual mode area.

Security Barriers on Highway

The vehicle is detected by infrared, and, being out of the dual mode area, does not require any radar confirmation. The radar detects the rail in the dual mode area, but IR does not confirm it, so it is discarded, therefore avoiding a false obstacle. Symmetric situations (IR and no radar) also happen.

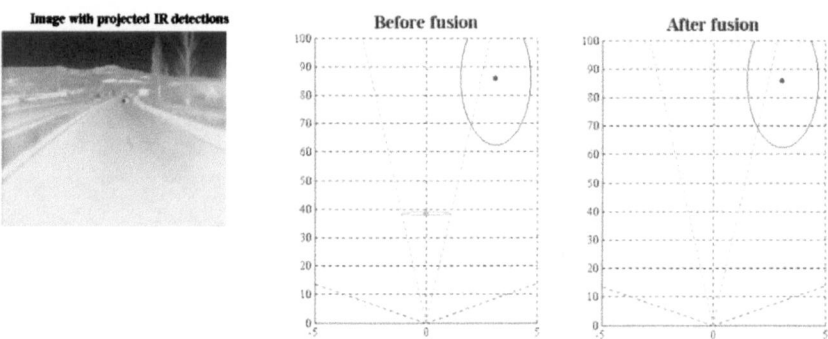

Fig. 17.   Radar detection not confirmed by infrared.

6   Driver's Interface

6.1   Visual Interface Design

Constraints linked with the driving task have been defined to make the interface more effective:

▶ Information given by the interface has to be detected without deviating the driver's gaze from the road and for that reason should be perceived in the driver's peripheral vision field.

▶ Information given by the interface has to be detected at any time, even if the driver is looking in another direction or performing a secondary task. In that case, flashing lights could be better detected.

▶ Information displayed has to indicate the obstacle approximate position quickly: blocks of LEDs could be dispatched all along the dashboard in front of the driver.

The visual interface is composed by 6 blocks (2 on the right, 2 on the left, 2 in the centre) displayed as follows:

Fig. 18. Visual interface.

6.2 Warning Modes

When it is required, signals are emitted through an interface to alert the driver. Two modes of warning can be displayed, mainly as a function of the time relationship between PAROTO car and a potential collision with the obstacle.

Attention Orienting Mode

When the situation is not of extreme urgency, the system will merely orient the driver visual attention to the obstacle direction. A set of amber flashing lights provided by LEDs will be displayed dynamically on the dashboard, in order to suggest the obstacle location and trajectory. In that case, the driver has time enough to take information from the environment and then, to optimize his or her response to a given situation. Such a warning makes possible to not disturb the driver with an intrusive alarm, when enough time is available. Moreover, it could increase the driver's confidence in the system, by allowing him to compare his perception of the driving situation to what the system is showing. It could also act in such a way that the driver becomes familiar with the system's information presentation. This could lead to a better understanding of the displayed information and then to quicker reaction in case of critical situations, which are very infrequent.

Alarm Mode

When the situation requires an urgent response from the driver, the alarm mode is activated. This mode, which is more intrusive, combines two types of displays: a visual display (red flashing LEDs located right in front of the driver) and an audio interface which produces sound. The alarm mode may be put into operation if the situation is very urgent or following an attention-orienting mode if the driver has not reacted.

6.3 Filtering

Various sensors on vehicle controls are also used to get information about what the driver is doing. Is he or she braking, turning the steering wheel, …? These sensors then inform the system whether or not the driver has seen an obstacle. For example, if the driver is well aware of the presence of an obstacle on the road and had already appropriately modified his or her behaviour to take it into account, it is not useful to provide him or her any alarm, due to the fact that alerting in this case, could lead to a decrease in the system acceptability. Designed using rule-based algorithms, the filtering module takes both fusion and sensor data into account.

7 PAROTO Platform

Infrared, radar and fusion modules run now in real time. Considering computational power needed by infrared and radar processing, we selected a distributed architecture, with one PC for each main task. Computers, running under different operating systems (which choice depends on specific needs), are connected to an IEEE 1394 bus and a standard LAN, thus allowing large band pass transfers.

Moreover, filtering module runs on fusion PC, making this computer connected to the CAN bus, on which circulates data from proprioceptive sensors. The filtering module controls the driver HMI through an RS-232 connection.

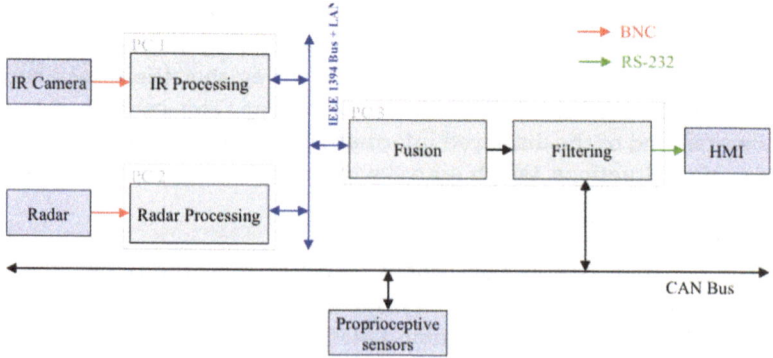

Fig. 19. PAROTO architecture.

8   Conclusion

We have presented the obstacle detection and avoidance function of the PARO-TO system.

An infrared camera and a radar yield detection tracks, passed to a fusion algorithm, which builds a precise description of the situation ahead of the vehicle.

Infrared, radar detection and fusion run in real-time on the testbed vehicle belonging to LASMEA This ambitious detection schemes has been thoroughly tested and proves very efficient in many different contexts.

The detected tracks are then passed on to a filtering process, which in turn decides whether it is pertinent to alert the driver, and the most appropriate way to do it, through a dedicated interface.

All these processing steps have been integrated successfully on the testbed vehicle, and are about ready for a thorough evaluation by INRETS experts in the coming weeks.

The major interest of all the technical achievements throughout the PAROTO project pushes us to investigate a low-cost industrial implementation.

References

[1]   L. Malaterre, "Système d'acquisition de données d'un radar automobile 77 GHz", Mémoire CNAM, France, mars 2003.

[2]   Y. Bar-Shalom T. E. Fortman, "Tracking and data association", Mathematics in Science and Engineering, 179.

[3]   J.B. Gao C.J. Harris, "Some remarks on Kalman filters for the multisensor fusion", Information Fusion 3, 2002, p.191-201.

[4]   S.S. Blackman, "Multiple-Target Tracking with Radar Application", ARTECH HOUSE.

[5]   Y. Le Guilloux, J. Lonnoy, R. Moreira, M.-P. Bruyas, A. Chapon, H. Tattegrain-Veste, "PAROTO Project: The Benefit of Infrared Imagery for Obstacle Avoidance", IEEE Intelligent Vehicle Symposium IV'2002, Versailles.

[6]   Y. Le Guilloux, J. Lonnoy, R. Moreira, "Infrared Image Processing for Obstacle Avoidance", AMAA 2003.

Y. Le Guilloux, J. Lonnoy, R. Moreira
SAGEM SA, Division Défense et Sécurité
95100 Argenteuil
France
yann.leguilloux@sagem.com

C. Blanc, J. Gallice, L. Trassoudaine
LASMEA, Laboratoire des Sciences et Matériaux pour l'Electronique, et
d'Automatique
UMR 6602 du CNRS
Université Blaise Pascal
Les Cézeaux 63177 AUBIERE Cedex

H. Tattegrain, M.-P. Bruyas, A. Chapon
INRETS, Institut National de Recherche sur les Transports et leur Securite
Laboratoire d'Ergonomie et de Sciences COgnitives pour les Transports
Case 24
25, av F. Mitterrand
69675 Bron cedex

Keywords: Infrared image processing, radar processing, data fusion, driver assistance, obstacle avoidance

From Passive to Active Safety Systems

T. Görnig, Conti Temic microelectronic GmbH

Abstract

State of the art automotive restraint systems have reached a very high level of occupant protection. Vehicle construction, belt and airbag systems are the most important lifesavers in this respect. The vehicle installation rate of restraint systems covers almost 100%. The development of passive safety systems is focused on an improved performance of the restraint devices while the need for the protection of all road traffic participants is increasing. This can be seen in upcoming regulations like pedestrian protection systems in the EU and the increasing activities in the area of active safety like TPMS in the US. This requires new types of sensors and system layouts. New technologies, material and processes will help to design and build new sensors with higher reliability and increased functionality.

1 Vehicle Safety Systems

Passive restraint systems have become standard equipment since the late 80'. We can find these systems in all vehicles that are on the road today. The systems are known as airbag systems and are state of the art of passive safety systems. Additional protection devices like foot bags, knee bags, inflatable belts and active headrest have also been introduced over the last years. All these systems have in common that the decision for activation of the restraint devices such as pretensioners, front airbags, side or head airbag is based on the signals from acceleration sensors. The accelerometers are installed in the airbag control unit and also in the side and upfront sensors that are used to sense the impact as early as possible.

The development of the restraint devices is following the general trend to more individually controlled devices as it can be seen on airbag modules. First systems used single stage inflators, the next generation was equipped with dual stage devices and latest developments are based on multi stage inflator technologies such as variable output inflators.

With the introduction of the different types of airbag systems also different type of crash sensors are required. For the first airbag systems with driver only or driver and passenger airbags a central crash sensing was required. The electronic modules where equipped with a single accelerometer or in cases where also an offset performance was required, two accelerometers in a ±45° orientation relative to the x-axis of the vehicle have been installed. This is the most common arrangement, which is still used in today's airbag control modules.

With the introduction of side airbags and head protection systems like Inflatable Curtains or Inflatable Tubular Structure additional side sensors were required. The first side sensors used an accelerometer and a micro controller for signal pre processing. The severity of the impact was transmitted as a PWM signal to the airbag control unit. Today's side sensor elements have a digital communication, transmitting the actual crash signal to the airbag control unit, where the signal is processed in the crash algorithm.

Up-Front sensors represent the latest level of innovation of a passive restraint system. They are installed close to the vehicle front and are used to detect a crash signal in the earliest possible phase of the impact. The purpose is a better discrimination of certain crash scenarios as well as to classify the severity of the crash to improve activation of smart airbag systems.

Some vehicle types like off road vehicles or SUV's use additional roll over sensors. The required components, gyros and low-g z-sensors are integrated into the airbag control module.

2 Crash Sensors

Early airbag systems used mechanical crash sensors. The sensors have been mounted on structural elements in the frontal area of the vehicle. The electrical contacts of this sensor type were closed on an impact and the airbags are inflated. A discrimination of the crash pulse was not possible with this type of sensor.

The next big step in the development of airbag control modules was the introduction of so called single point sensing systems. This type of systems became possible with the introduction of electronic crash sensors. The early generation of sensors have been piezoelectric sensors, followed by bulk micro machined resistive sensors and surface micro machined capacitive sensors. An improved discrimination of the crash pulse became possible with the development and introduction of this type of sensors.

The state of the art crash sensors consists mostly of a micro machined sensing element and an ASIC that is used for signal conditioning, amplification and filtering. The ASIC also includes an interface that serves as a connection to the control module. The sensor is in most cases supplied with the required operating voltage by the control module and the signals from the sensor to the control module are transmitted by current modulation. Figure 1 shows a typical two chip crash sensor assembly and the block diagram.

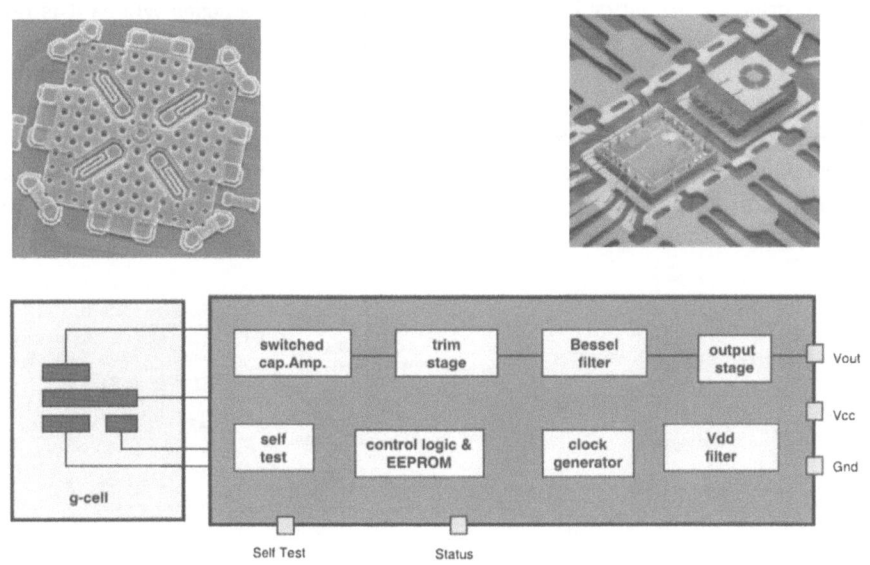

Fig. 1. Sensor assembly.

When we look at the future of the safety systems that we know today, the passive safety systems, we have to realize that the penetration of passive airbag systems in the market has reached a relative saturation. The major potential can be seen in the further development of smart airbag systems. Also a shift from passive safety towards active safety can be seen on the market. This is obvious with the increased number of systems that are used for improved braking like ABS, BAS or vehicle dynamics control systems like ESP and roll over prevention systems. A further development towards active safety systems are advanced ACC systems, Stop & Roll and Stop & Go systems.

Future active safety system applications are e.g. lane detection, lane keeping support, driver warning and driver assistance systems, blind spot detection systems, crash avoidance and automatic cruising systems, which we might see in the future.

3   Future Safety Systems

From the actual road traffic and accident statistics it is known that passive safety systems have proven be very efficient in the reduction of injuries in vehicle crashes. Statistics it is also shows that there are cases where it is necessary to improve the performance of passive safety systems.

This data also shows that protection for other road traffic participants is required, as there are motorbike drivers, bike riders and a large group of pedestrians. This can be seen in the regulation efforts regarding pedestrian protection systems of the European Union and the self-commitment of the European Automobile Manufacturer Association ACEA.

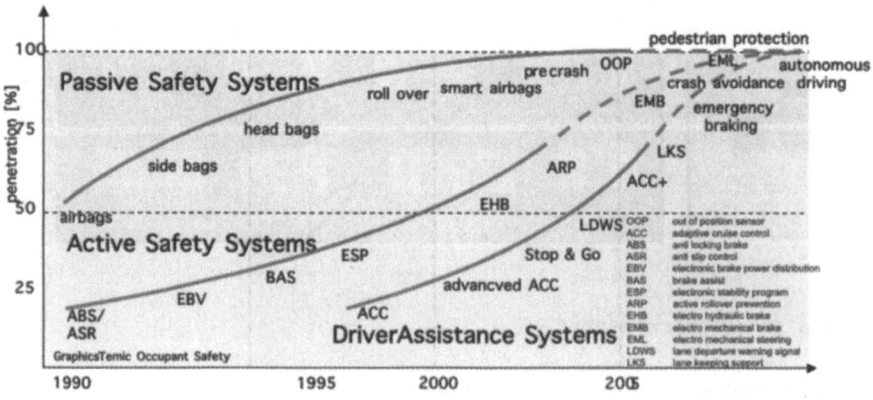

Fig. 2.   Market trends.

Passive safety systems are activated at the moment when the impact occurs. The purpose of passive safety systems is to offer the best possible protection against injuries for the vehicle occupants, to limit the physical load of the occupant. Due to this limitation a new generation of sensor and sensing systems becomes necessary to fulfil the requirements of a future passive safety system.

For a future safety system it is required to get information about a possible impact before the actual impact occurs. Depending on the range of the sensor, the information is used for driver warning, for driver assistance and finally for crash avoidance. The signals will also be used for activation of a restraint system with a much higher precision and a better restraint performance compared to today's systems. To generate reliable information before the possible impact, it is required that the sensor measures the proximity of objects within the complete area of a vehicle.

4   Pre-Crash Sensor

A first generation of a pre-crash sensor consists of an optical sensor system that is mounted behind the windscreen, in a central position close to the rear mirror in the wiper area. The sensor principle is based on the speed of light. Infrared coded laser light is sent out and reflected from objects in the proximity of the sensor. The decoder circuit calculates the time of flight between the emitted and received laser pulse. A wide beam arrangement has been selected for the sensor system to cover the complete area in front of the vehicle.

- Pre-Crash Sensor

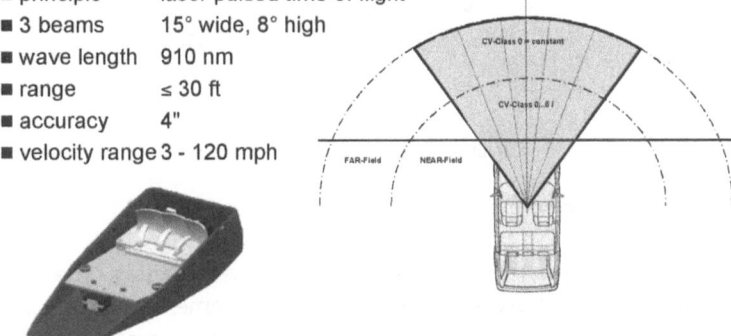

■ principle	laser pulsed time of flight
■ 3 beams	15° wide, 8° high
■ wave length	910 nm
■ range	≤ 30 ft
■ accuracy	4"
■ velocity range	3 - 120 mph

Fig. 3.   CV sensor.

The primary application of the CV Sensor is the usage as a pre-crash sensor. The sensor system measures the distance of an object in the front of the vehicle. The speed of an approaching object is calculated from the distance information of repetitive measurements. With the wide beam and multiple receiver arrangement it also possible to calculate the direction of the approaching object.

With this information it is very early known, before the actual impact occurs, which restraint device should be activated. It is also possible to estimate which severeness the impact will have since the relative impact speed is calculated from the sensor signals. This information is used to control restraint devices like adaptive force limiters or adaptive airbag systems as well as reversible restraint devices.

Due to the fact that the relative impact speed can be calculated very precisely, the activation of restraint devices can also be staged much more efficient.

A further advantage of the sensor arrangement is that the information is independent from the vehicle structure, variations or ageing effects that might influence the crash performance of a conventional sensing system are eliminated.

The key component for the pre-crash sensor is a custom designed laser diode that consists of a laser chip the necessary power switch, a MOSFET, and the energy reserve which consists of two ceramic capacitors.

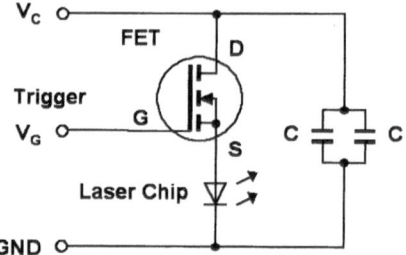

Fig. 4.   Circuit diagram.

This optoelectronic micro system is a lead frame assembly. The components for the energy reserve are soldered to the lead frame, the laser chip and the MOSFET are attached as single chips. The connections are conventional bond technology, the complete device is moulded in a clear plastic material.

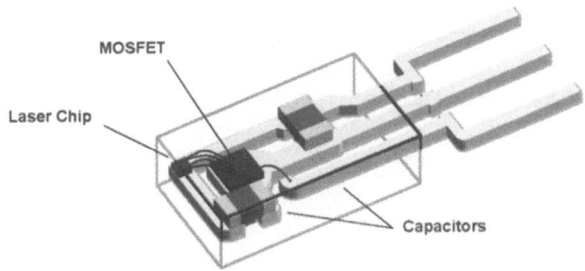

MOSFET

Laser Chip

Capacitors

Fig. 5.   Circuit diagram.

A further application of the pre-crash sensor is to use it for an active pedestrian protection system. The introduction in Europe is planned from September 2005 on. The first systems will include an active hood, which is lifted for a couple of centimetres to give additional deformation area. The next generation, that are introduced from 2010 on will include airbags at the a-pillar or the lower windscreen frame to protect the pedestrians from direct contacts with hard points of the vehicles.

The purpose of the sensor is to detect an object, in this case a pedestrian in the vicinity of the vehicle and to predict the impact time and speed of the object. In parallel a measurement window is opened to receive a confirmation of the physical impact from an additional force sensor or bumper mounted accelerometers. With its capability to track objects after the confirmed physical impact, "hard" and "soft" objects can be discriminated. This additional information is used for discriminate a pedestrian impact from an ordinary crash. The algorithm that is operated in the control unit will activate the necessary protection devices.

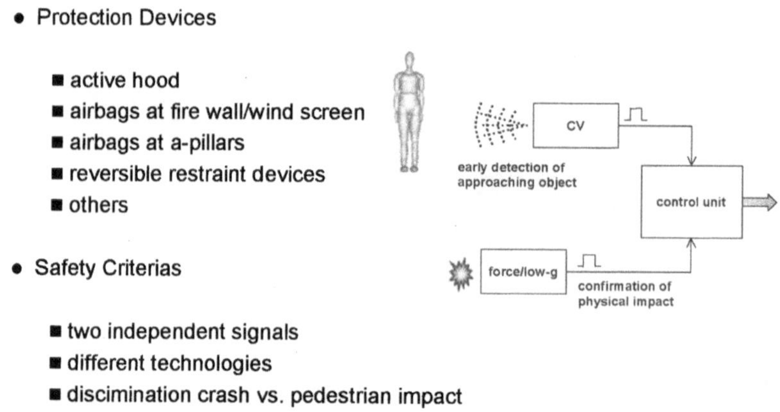

- Protection Devices

  - active hood
  - airbags at fire wall/wind screen
  - airbags at a-pillars
  - reversible restraint devices
  - others

- Safety Criterias

  - two independent signals
  - different technologies
  - discimination crash vs. pedestrian impact

early detection of approaching object

CV

control unit

force/low-g

confirmation of physical impact

Fig. 6.   Pedestrian protection system.

5   Active/Passive Safety System

The trend towards improved passive safety systems puts also new demands to the design of restraint systems. Due to the increased number of sensors the requirement to build up sensor clusters is obvious. Another important issue is the wiring effort for this type of systems. It has to be reduced for costs, assembly, weight, and quality and reliability reasons.

Bus systems will be used for the connection and transmission of information between the different modules. There will be different types of bus systems, depending on the different applications that we can find for future restraint systems in vehicles. There will be high sensor bus systems as they are required for transmitting crash sensor information and there will also be sensor bus systems for smart peripheral devices. This bus system is used to transmit the information from buckle switches or seat track sensors to a restraint control module as well as to transmit information to actuators, e.g. reversible pretensioners. The Bosch/Siemens/Temic Sensor and Deployment Bus describe a capable bus system. Related bus compatible igniters for actuation of passive restraint devices are developed in the cooperation project IBA.

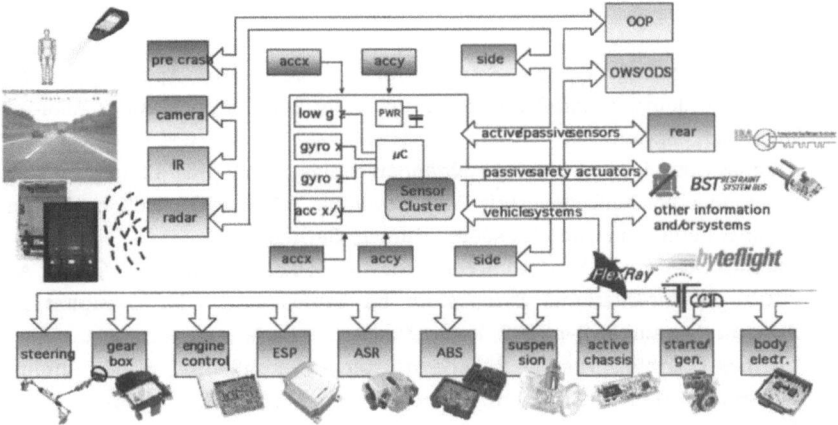

Fig. 7.   Active/passive system integration.

6   Conclusion

Future restraint systems require new types of sensors and also new technologies for the sensors. They will require pre-crash sensors with different sensing ranges, resolutions, wide and narrow focused beams. Their will also no longer be single sensor arrangements for future restraint systems. pre-crash sensors will also help to increase safety for all road traffic participants significantly. The challenge for the future is to find the best possible combination of the different available sensors and technologies to get a maximum of performance at economical costs.

References

[1]   Thomas Goernig: Optical Sensors for Passive Safety Applications Bayern Innovativ, Jahreskongreß Zulieferer Innovativ 2002, July 3 2002, Audi-Forum, Ingolstadt, Germany

[2]   Thomas Goernig: Sensoric für Fußgängerschutzsysteme, HdT Tagung - Schutz der Fußgänger bei Kollisionen mit Pkw, November 20-21 2002, HdT im Forum der Technik beim Deutschen Museum, Munich, Germany

[3]   Thomas Goernig: "Advanced Sensors for Future Restraint Systems", V10, Airbag 2002, 6th International Symposium on Sophisticated Car Occupant Safety Systems, December 2-4 2002, Kongreßzentrum, Karlsruhe, Germany

[4]   Thomas Goernig: "Passive Rückhaltesysteme der Zukunft", p 88, Automotive Electronics, März 2003, Friedr. Vieweg & Sohn Verlag, Wiesbaden

[5]   Thomas Goernig: "pre-crash-Sensierung / Fußgängerschutz", p 81, BAIS-Frühjahrstagung - Neue Aspekte der Fahrzeugsicherheit bei PKW und Krad, March 28-30 2003, Nürnberg, Germany

Thomas Görnig
Conti Temic microelectronic GmbH
Ringlerstraße 17
85057 Ingolstadt
Germany
thomas.goernig@temic.com

Keywords: Sensors, Microsystems, Optical Sensors, IBA, Bus Systems

A Low-g 3 Axis Accelerometer for Emerging Automotive Applications

B. Grieco, D. Ausilio, F. Banfi, E. Chiesa, G. Frezza, E. Lasalandra,
P. Galletta, M. Garavaglia, A. Mancaniello, A. Merassi, F. Pasolini, G. Sironi,
F. Speroni, M. Tronconi, T. Ungaretti, B. Vigna, S. Zerbini, ST Microelectronics

Abstract

A 30 µg/sqrt(Hz) resolution MEMS based low-g 3-axis accelerometers targeted for emerging automotive application is presented. The device is composed by 2-dice in a single package: the sensing element and the processing IC. The sensing element is realized with an epitaxial micro-machining process (ThELMA™) while the IC uses a 0.5 µm BICMOS process. The IC amplifier is able to detect capacitive changes as low as 10 aF ($10 \cdot 10^{-18}$ F) and the device is able to provide both an analog and a digital output. Moreover it has very good linearity and very low cross talk between the sensing axes thanks to a design that fits very well the ST micro-machining process ThELMA. Finally, the proposed interface accelerometer is factory-trimmed to ensure repeatable performance without production-line adjustments in the end product

The device is intended for new inertial applications in the automotive field that are expected to be heavily dominated by micro-machined silicon components More specifically multiaxis accelerometers are particularly useful for emerging application like new generation ESP and ABS system, assisted navigation, headlight leveling, car alarm and virtual horizon.

1 Introduction

LIS3L02 is the first 3 axis low-g linear accelerometer on the market able to measure 30 millionths (micro) of gravity acceleration with the lowest power consumption at 2.7 Volt supply integrated in a single package. The output data are available in analog and digital format. The device is also very robust against mechanical shock being compliant with drop test requirements.

LIS3L02 is a dual silicon chip solution inside a single package (PSO24: 15.3 mm x 7 mm x 2.5 mm or QFPN44: 7 mm x 7 mm x 1.8 mm): the sensor chip and the interface electronics. This approach allows the maximum flexibility, genericity and modularity together with a shorter development time for new products.

Most advanced solution currently available on the market can provide only:
▶ Maximum of 2-axis sensing feature integrated with the electronics in a single package.
▶ 3 axis sensing element without the interface electronics.
▶ Bulky solutions at PCB level realized by assembling 1 or 2 axis accelerometers.

Among all the competing 3 axis accelerometers only ST 3-axis accelerometer is able to provide outstanding performances in a small package with low power consumption.

The achievement of this breakthrough in the field of micro-machined accelerometers has been possible thanks to different innovations, which allowed overcoming the following technical challenges:
▶ Achievement of an excellent process control in the manufacturing of an in-plane (XY) and out-of-plane (Z) sensitive micro-mechanical structure on the same die. The challenge was to optimize and control both the gap and the thickness of the sensitive element.
▶ Development of two processing steps (Epitaxial Growth and Grinding technology) able to realize thick (15 µm) structures with low mechanical stress. This innovation step allowed reaching high resolution and high sensitivity without using the expensive Silicon-On-Insulator wafer and the difficult to industrialize bulk micro-machining technology.
▶ The high robustness of the device has been obtained through an innovative mechanical design and the development of a high yield strength epitaxial poly-silicon layer (~ 5 GPa).
▶ Two level polysilicon layers and the high aspect ratio silicon etching technology allows to measure accelerations along the desired axis with low cross talk among the different axis.
▶ Thinning of the silicon die and stacking of the sensing element and IC allowed the realization of such small package format
▶ An innovative electronic circuitry architecture together with an accurate layout resulted in an excellent resolution for all the 3 axis with the small voltage supply and the low power consumption (less than 2 mW).

2   Fabrication Process: ThELMA™ Technology

STMicroelectronics has developed a technological process for the realization of a complete family of inertial sensors (angular, linear accelerometers and gyroscopes), This process is called ThELMA and it stands for Thick Epitaxial Layer for Microactuators and Accelerometers (see figure1). On a standard silicon substrate a thermal oxide is grown. On the top of the oxide a thin polysilicon layer is LPCVD deposited in order to realize buried interconnections. Afterwards a sacrificial oxide layer is deposed on the top of which a thick polysilicon epithaxy is grown (15 mm). This epithaxial polysilicon is the structural material that composes the sensor. The structure of the sensor is defined by deep silicon RIE. At the end the sacrificial oxide is removed by using a vapor phase HF etching.

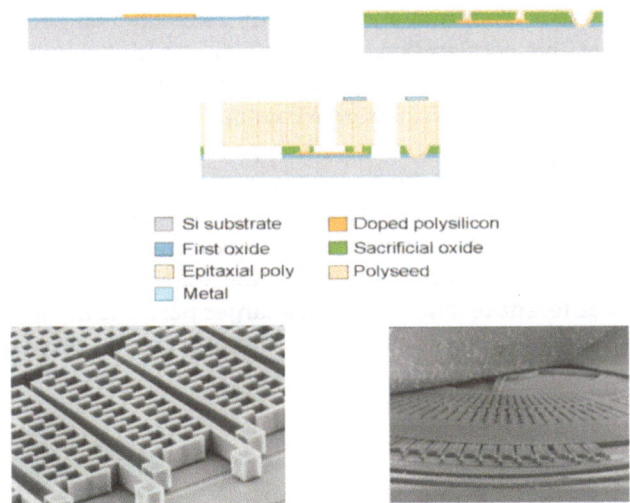

Fig. 1.   ST ThELMA micro-machining: Schematic process description and scanning electron microscope pictures.

With the process described above the sensor results to be thicker than sensors obtained with thin film deposition techniques and therefore shows an higher sensitivity; moreover with respect to the structures obtained with surface micromachining technique it has greater stiffness to bending along z-axis and therefore a lower probability to stick on the substrate during sacrificial oxide removal. This process has also a lower cost than processes based on the use of SOI or silicon fusion bonded substrates.

Due to the necessity to protect the devices during wafer sawing process and to guarantee a controlled atmosphere during their lifetime, a process for encapsulating the sensors at wafer level has been developed. The process consists of using another silicon wafer to create the caps for the sensing structure with holes trough it in correspondence to pad region and of using an hermetic wafer to wafer bonding technique to seal the device as shown in the figures below.

3   Chip Partitioning and Interface Electronics

The device is composed by a triaxial sensing element and by an electronic front-end. which has been fully integrated in a 0.5 μm standard CMOS technology. The overall system exhibits a 30/sqrt(Hz) · μg noise level and a full scale of 2 g or 6 g that can be selected by supplying the right voltage to a specified pin of the component. The front-end is directly connected on package to a tri-axial linear capacitive acceleration-sensing element and it supplies an analog voltage (V/g) or digital (SPI) output proportional to the input acceleration. The IC has the capability of processing signal coming from a sensor not realized in the same wafer, but connected in the same package. This feature allows the optimization of the sensor and the electronic front-end performance since they can be realized in different, and, then, customized and more suitable technologies. The use of a single technology typically results in stringent trade-off in sensor and electronics performance. On the other hand, the cost of the choice of using different technologies is the larger parasitic input capacitance, which requires higher power consumption to reach the same dynamic range.

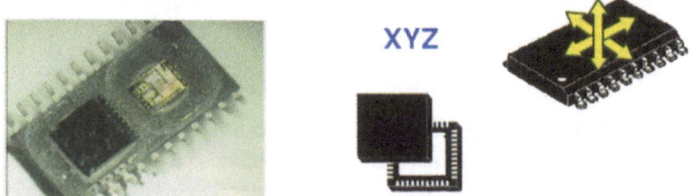

Fig. 2.   Encapsulated Sensing element and IC are shown on the left; the complete PSO24 package (15 mm x 7 mm x 2.5 mm) and QFPN44 (7 mm x 7 mm x 1.8 mm) on the right.

The overall scheme of the proposed analog tri-axial linear accelerometer is shown more in detail in figure 3. A capacitive MEMS-based linear acceleration sensor is mounted, where a single mass is available to sense both x-axis, y-axis and z-axis acceleration.

The three sensing axes are intrinsically aligned precisely at 90° to each other. As a result, the board mounting is much simpler than when two separate 2-axis linear sensors are used. In the sensor structure, a moving mass is present, while two stators for each axis are used (S_{1x} and S_{2x} for the x-axis, and S_{1y} and S_{2y} for the y-axis, S_{1z} and S_{2z} for the z-axis).

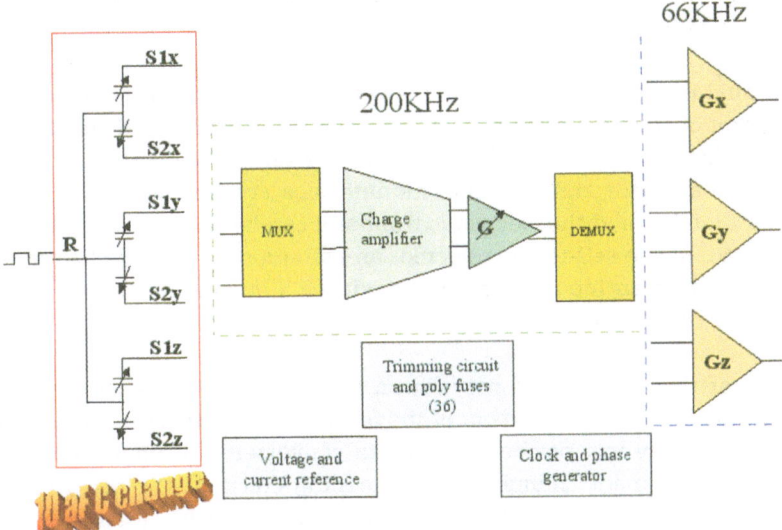

Fig. 3. Sensing element in red block and IC on the right side of the picture. In case of the digital output device a proper ADC is present as an additional block of the above picture.

The true sensor capacitances (whose variations have to be measured) are Cs_{1x} and Cs_{2x} for the x-axis, Cs_{1y} and Cs_{2y} for the y-axis and Cs_{1z} and Cs_{2z} for the z-axis. When a linear acceleration is applied to the sensor, the proof mass displaces from its nominal position, causing an imbalance in the sensor capacitive half-bridges of Cs_{1x} and Cs_{2x} for the x-axis, and of $Cs_{1y,z}$ and $Cs_{2y,z}$ for the y,z-axis. The interface chip translates these minimal capacitance changes into calibrated analog voltages at the output pin proportional to the proof mass movement, and so to the applied acceleration.

The capacitive imbalance is measured using charge integration in response to a fixed voltage pulse applied to the sense capacitors. In this way the interface front-end structure can be implemented by using differential switched-capacitor techniques, which have demonstrated to be able to perform high dynamic range analog circuits. In addition, the complete signal processing uses a fully differential structure to improve system performance robustness. The multiplex (MUX) and the demultiplex (DEMUX) techniques in the scheme are implemented in the digital part controlling the switch and not using additional series switch which could be detrimental for the system performance. The offset and $1/f$ noise of the IC are cancelled through the only-passive Correlated-Double-Sampling (CDS) structure implemented at its output nodes. This solution has been preferred to other ones implemented at the opamp input nodes due to its higher performance robustness with respect to parasitic capacitance and charge injection.

As a final feature, a self-test mode is available, which is based on electrostatic actuation of the acceleration sensor proof-mass. Through a dedicated self-test pin is possible to electrostatically deflect the sensor beam at any time to verify that the sensor and its electronic interface are functioning correctly. The sensor mobility and the electronic front-end can be checked using the self-test mode. When the self-test is selected, an actuation force is applied to the sensor simulating an input force. This produces a change in the DC level of the output of the device.

Moreover the IC has some poly-silicon fuses to trim the offset and the sensitivity of the sensors coming out of the fabrication line.In case of the digital output device very low power Sigma-Delta converters are implemented for the three measurement channels, while keeping constant the overall resolution and power consumption thanks to smart tricks in the IC design.

4   Measurements Results

4.1   Sensing Element Transfer Function

To achieve the high performances of 3-axis accelerometer is is extremely important to control very well the mechanical parameters of the sensing element: the resonance frequency and the quality factor. The sensing masses have resonances frequeny between 2 and 4 kHz while the damping factor ranges from 0.5 for z-sensing element to 2 for xy-sensing element. Here below the reader can find the dynamic transfer function of the sensing element:

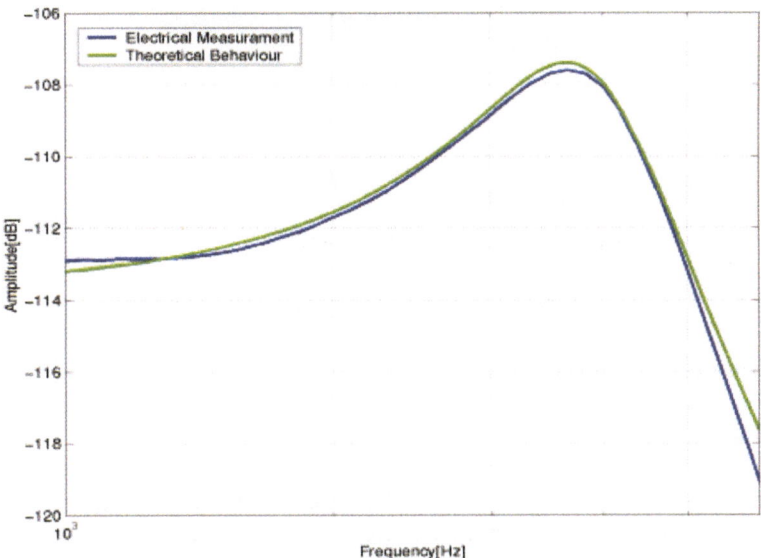

Fig. 4.   Mechanical trasnsfer function of the z-axis sensing element and x/y sensing elements.

4.2   Linearity of the 3-Axis Accelerometer

Testing was performed in 5 different temperatures using a temperature chamber: +85°C, +55°C, +20°, -25°C and -40°C. The units were turned 90° in 10° steps starting from the angle where the gravity effect to the sensor was 0, using a rotary table. After 90° turn the gravity effect to the sensor was maximum (1 g).

$$g_{vertical} = \sin(\alpha) * g_n$$

The output voltage can be presented as a function of the normal acceleration. A best fit linear trend line is fitted to the results and error from the measured results to the fitted line is calculated. The result is average percentage part of 1 g scale. All the results are definitely below 1% of the product specs.

Temperature (°C)	Non-linearity (%)
+85	0,40
+55	0,25
+20	0,10
+20	0,32
-25	0,13
-40	0,18

Table 1. Linearity of the 3-Axis Accelerometer.

5 Temperature Measurements

5.1 Sensitivity Stability

Sensitivity Stability over temperature has been measured and the results show that the sensitiviy has a thermal coefficient of less than 0.02%/degree in the temperature range from -40°C to +85°C.

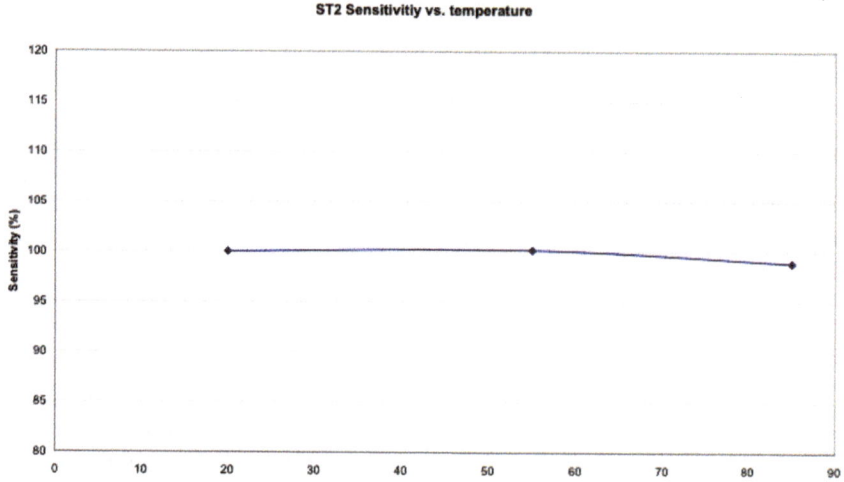

Fig. 5. Thermal stability of the accelerometer sensitivity.

5.2 Offset Stability

The package of our device is full molded and therefore the zero-g offset is more sensitive to the temperature effect. However besides the measurements reported below we are studying new mechanical designs and a new IC aimed to improve the thermal offset stability of the 3-axis accelerometer to match the strong requirements of some specific automotive applications.

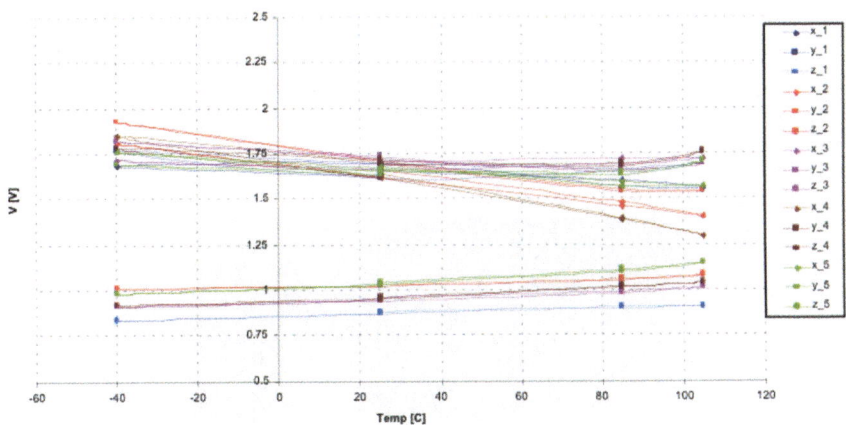

Fig. 6.   Thermal stability of the zero-g offset. The device is within ±10%.

5.3 Cross Axis Sensitivity

The perfect alignment guaranteed by the photo-masking step together with a good mechanical design allow to optimize each sensing element for each sensitive axis. Therefore the cross axis sensitivity is lower than 1% while products based on bulk micro-machining are able to guarantee only 4-5% of cross axis sensitivity. This feature is extremely important for all the applications where the spurious accelerations must be rejected.

5.4 Complete Measurement

The following graphs show the analog output and digital output of the 3-axis accelerometer from ST Microelectronics.

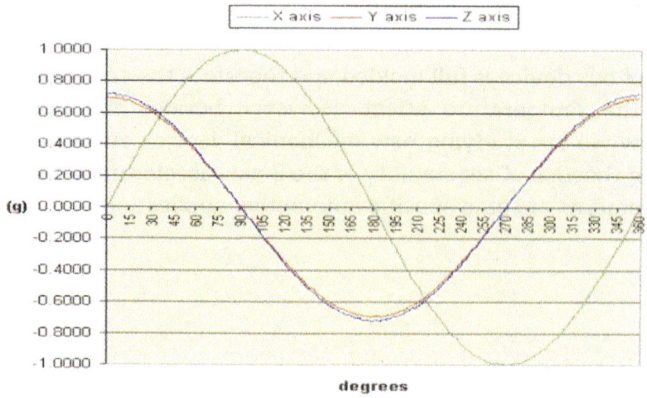

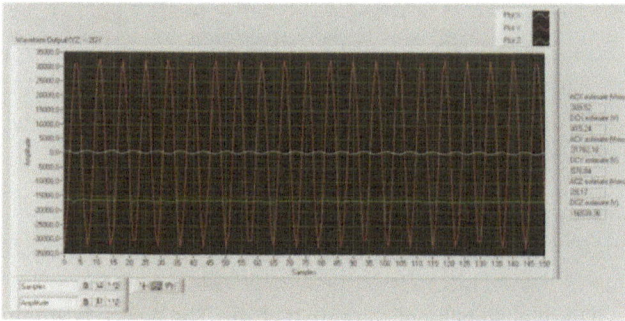

Fig. 7.   Analog and digital output graphs of 3-axis accelerometer. While the first graph show results in V/g the second has a number associated with the input acceleration.

6   Conclusions

We have developed a low power consumption (< 2 mW), high sensitive, 3-axis accelerometers that could be used for many emerging applications in the car.

The device has an analog and a digital output and a bandwidth quite large. The stability of the sensitivity over the temperature range is very good together with the linearity and the cross axis sensitivity (< 1%).

Some features need still to be improved for some automotive applications, but the first results currently available on new mechanical structures in different packages show very good promises for further improvements of the devices while always keeping the power consumption very limited and the sensitivity very good.

Acknowledgements

The authors would like to acknowledge all the ST technological development group led by Dr. P.Ferrari for the tremendous amount of work they spend to fix the micromachining THELMA process. Morevoer a special thanks to Bruno Repellini, Alessandro Vinay and Biagio De Masi to have conceived all the test structures and to have implemented all the proper routines to keep under control the ThELMATM process.

References

[1]   A. Garcia-Valenzuela and M. Tabib-Azar, "Comparative study of piezo-electric, piezoresistive, electrostatic, magnetic, and optical sensors," Proc. SPIE, July 1994, vol. 2291, pp. 125–142.

[2]   W. Yun, "A surface micromachined accelerometer with integrated CMOS detection circuitry," doctoral dissertation, Univ. of California at Berkeley, 1992.

[3]   W. Gopel, J. Hesse, and J. N. Zemel, Eds., Sensors: A Comprehensive Survey. Weinheim, Germany: Wiley, vol. 7, 1994.

[4]   S. J. Sherman, W. K. Tsang, T. A. Core, and R. S. Payne, "A low cost monolithic accelerometer; Product/technology update," in Proc. IEDM, San Francisco, CA, Dec. 1992, pp. 501–504.

[5]   C. Lu, M. Lemkin, and B. E. Boser, "A monolithic surface microma-chined accelerometer with digital output," in ISSCC Dig. Tech. Papers, Feb. 1995, pp. 160–161.

[6]   Gabrielson, T. B., Mechanical-Thermal Noise in Micromachined Acoustic and Vibration Sensors: IEEE Transactions on Electron Devices, Vol. 40, No. 5, May 1993, 903-909.

[7]   E.M.A Pleska, J.F. Keiffer, P. Bouniol, Les senseurs inertiels du XXIe siecle, RSTD No 49 – STTC – Juillet 2000.

[8]   A. Lawrence, Modern Inertial Technology, Second Edition, Springer.

[9]   L. J. Ristic, R. Gutteridge, B. Dunn, D. Mietus, and P. Bennett, "Surface microma-chined polysilicon accelerometer," in Proc. IEEE Solid-State Sensor and Actuator Workshop, Hilton Head Island, SC, June 1992, pp. 118–121.

[10] W. Henrion, L. DiSanza, M. Ip, S. Terry, and H. Jerman, "Wide dynamic range direct digital accelerometer," in Proc. IEEE Solid-State Sensor and Actuator Workshop, Hilton Head Island, SC, June 1990, pp. 153–157.

[11] W. A. Clark and R. T. Howe, "Surface micromachined z -axis vibra-tory rate gyro-scope," in Proc. IEEE Solid-State Sensor and Actuator Workshop, Hilton Head Island, SC, June 1996, pp. 283–287.

[12] C. J. Kemp and L. Spangler, "An accelerometer interface circuit," in CICC Dig. Tech. Papers, Santa Clara, CA, May 1995, pp. 345–348.

[13] E. Obermeier, "Processes for the fabrication of Three Dimensional MEMS structures", AMAA 2003

[14] T. Riepl et al., "Signal Procesign for Automotie Applications", AMAA 2003

[15] E. Jung et al., " Packaging of Micro Devices for Automotive Applications – Techniques and Examples", AMAA 2003

[16] K. Kapser et al., "A low-g Accelerometer for Inertial Measurement Units", AMAA 2003

[17] M. Lemkin, Micro Accelerometer Design with Digital Feedback Control, doctoral dissertation, Univ. of California at Berkeley, 1997.

[18] B. E. Boser and R. T. Howe, "Surface micromachined accelerometers," IEEE J. Solid-State Circuits, vol. 31, pp. 366–375, Mar. 1996.

[19] T. C. Choi, R. T. Kaneshiro, R. W. Brodersen, P. R. Gray, W. B. Jett, and M. Wilcox, "High-frequency CMOS switched-capacitor filters for communications application," IEEE J. Solid-State Circuits, vol. SC-18, pp. 652–664, Dec. 1983.

[20] J. H. Smith, S. Montague, J. J. Sniegowski, J. R. Murray, and P. J. McWhorter, "Embedded micromechanical devices for the monolithic integration of MEMS with CMOS," in Proc. IEDM, Dec. 1995, pp. 609–612.

[21] T. Cho and P. Gray, "A 10 b 20 Msample/s, 35 mW pipeline A/D converter," IEEE J. Solid-State Circuits, vol. 30, pp. 166–172, Mar. 1995.

[22] C.Hernden, "Vibration cancellation using rotational accelerometer feedforward in HDDs" Data Storage (November 2000).

[23] P.Adrian, " Sensor Business, Marketing and Technology Developments", Sensor Business Digest, Nov 2003.

[24] Strategy Analitics, "Automotive Silicon Sensor Market", February 2003

[25] www.st.com, LIS3L02AS and LIS3L02AQ Datasheets and Application note

Barbara Grieco, Benedetto Vigna
ST Microelectronics
Via Tolomeo, 1
20010 Cornaredo
Milano
Italy
Barbara.Grieco@st.com, Benedetto.Vigna@st.com

Keywords: Multi-axis accelerometer, high preformance accelerometer, Epitaxial Micro-Machining, Full-molded accelerometer, low-power consumption accelerometer, system in Package, Digital output accelerometer

Powertrain

Autonomous Sensor Systems for Car Applications

A. Bodensohn, R. Falsett, M. Haueis, M. Pulvermüller, DaimlerChrysler AG

Abstract

The development of new powertrain functions has many goals: more power and torque, low fuel consumption and emission, best driving performance and a brand specific acoustics. In the future the accuracy of control in Motor Control Units (MCU) will increase more and more. For this reason open loop control will replaced by closed loop control. New and/or other values are needed for better algorithms in powertrain control. Autonomous sensor systems for automobile industry offer increased reliability due to the reduction of cable connections, novel sensing applications made possible by wireless data transfer and cost advantages resulting from the aforementioned aspects. These systems consist of a power generator, energy storage and data transceiver. The technological challenge for realizing such system is most of all the construction and fabrication of a mini power generator that matches the dimensions of standard sensor modules. In this paper different power generators for autonomous sensor systems are reviewed and systematically discussed. It is shown that low power thermoelectric generators can supply autonomous sensor systems with up to 7 mW utilizing waste energy of the engine.

1 Power Generation in Autonomous Sensor Systems

Condition monitoring systems in nowadays cars apply more than 120 sensors distributed throughout the body of the car. Wires to connect them impose many limitations, especially if the sensor is located in a harsh environment or a remote location. Sensor-to-controller cables bring disadvantages to the cost structure, the weight and therefore emissions of the automobile and the reliability of the sensor system. Autonomous sensor systems promise a new generation of wireless sensor networks for automotive industry. Autonomous sensor systems will reduce the number of cable connections, will make novel sensing applications possible by wireless data transfer and will result in valuable cost advantages.

A wireless sensor system consists of a power generator, energy storage and data transceiver (figure 1).

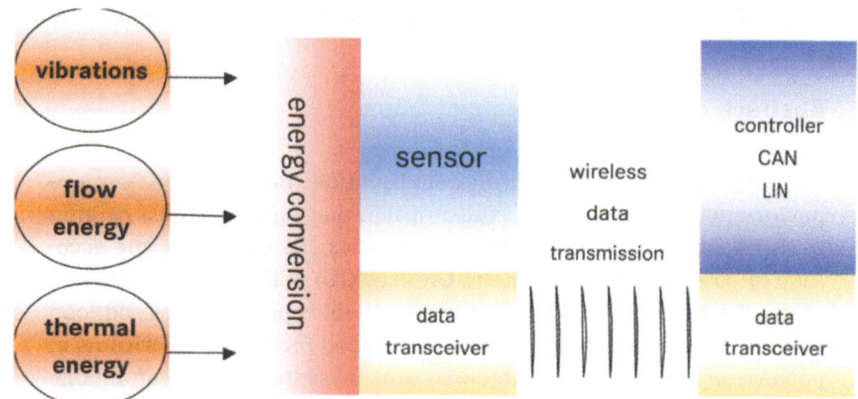

Fig. 1.   Graphical illustration of a wireless sensor, possible energy sources and its communication with external control units.

The technological challenge for realizing such system is most of all the construction and fabrication of a power generator that is small enough to match the dimensions of standard sensor modules. Micro- and Nanotechnology are privileged to solve such challenges and power generators with output of several microwatts were presented in literature [1-3]. Earlier work on piezoelectric, electrostatic and electromagnetic vibrational energy conversion schemes illustrate the great interest in micro power generators for applications where batteries are unacceptable [5-7]. Looking at the energy budget of an automobile someone finds that only roughly 25% of the energy are used for mobility and accessories. Approximately 75% of the energy supplied for combustion are lost as exhaust gases, coolants, friction and radiated energy. The conversion of thermal and kinetic energy into electricity is consequently the preferred principle for energy supply in autonomous sensor systems for car applications [4]. Kinetic energy conversion generated from vibrations was described in detail by Hettich et. al. [4]. The vibration response of the chassis of a Mercedes-Benz M-Class was measured for different road conditions. The typical peak acceleration, for example for driving on paving stones (30 km/h) is 40 m/s^2. An inductive power generator as demonstrated by Hettich et. al. supplied a peak power of 10 mW whereas the average power of the generator is almost one order lower. The typical power range of automotive sensors is several milliwatt power, which is significantly more than what many of the above mentioned demonstrators are designed for. The energy density of usable kinetic energy compared to the energy density of heat in a car is about two orders lower [7]. Especially sensor applications in the powertrain and the exhaust system could greatly benefit form thermoelectric power generators.

For data transmission with a radio frequency interface we use well known modules from commercial providers. Different carrier frequencies and protocols such as a bluetooth module, WLAN module or a 433 MHz or 866 MHz transceiver are available. Since they are described by suppliers they are not part of this paper. The choice of the appropriate transmitter depends on the transmitted data rate, interfaces and environment.

For example for temperature sensors in automobiles a low data rate readout is sufficient since they do not change the value relative often. In many applications it is not necessary to read the sensor output more then one time per second. A RF data transmission takes only a few milliseconds. The remaining time, the electronic is in sleep mode and does not consumes energy. For this kind of sensors, without high timing requirements, we can use advantageously autonomous sensor systems with power generation.

2   The Thermoelectric Generator

This paper describes a new approach for power generation from thermal energy in the powertrain of a car. A low power thermoelectric generator (LPTG) was developed and realized in cooperation with D.T.S. GmbH, Halle. The LPTG was manufactured with thin film technology methods and equipment for batch processes for film generation, photolithography, pattern generation, wafer and foil cutting and device assembly [8]. This approach is different from standard thick film technologies offered by most other commercial suppliers of thermoelectric generators and is new to automobile applications. The demonstrated LPTG-prototype consists of 20475 thermocouples made of a doped bismuth telluride alloy (p-leg: $(Bi_{0.25}Te_{0.75})_2Te_3$, n-leg: $Bi_2(Se_{0.1}Te_{0.9})_3$). The active films have a thickness of 2.5 micrometer, the internal electrical resistance amounts 3.8 kOhm, see figure 2. The volume of the packaged thermoelectric generator is 0.5 cm^3.

The figure of merit for thermoelectric generators, denoted ZT, is given by:

$$ZT = \frac{S^2 \cdot \sigma \cdot T}{\kappa} \tag{1}$$

where S is the Seebeck Coefficient, s the conductivity, T the temperature and κ the thermal conductivity. The figure of merit is an expression for the efficiency of converting thermal energy into electricity. The described thermoelectric generator has a Seebeck Coefficient of 0,26 V/K and a thermal conductivity of 0,161 W/K. The focus of design optimization has been maximum power density. This has been achieved at the cost of reduced efficiency. Commonly used thermoelectric generators are made of doped bismuth telluride alloys in bulk technology and achieve an efficiency as high as 6%. The described thin film devices, however, only achieved an efficiency of about 1%.

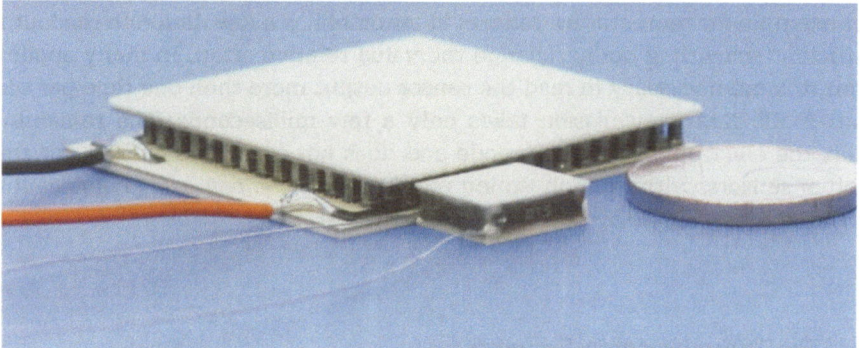

Fig. 2.   Photograph of a thermoelectric generator using foil technology (LPTG) and a conventional thermoelectric generator using standard bulk technology.

3   Characterization of the Thermoelectric Generator in an Engine Environment

The thermoelectric generator was tested under conditions as they are typical in the vicinity of the engine. For that purpose an oilpan of a 320 CDI diesel engine was modified so that a thermoelectric generator and cooling fins could be mounted (figure 3).

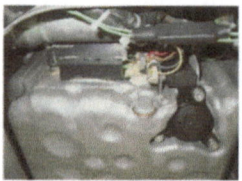

Fig. 3.   Thermoelectric generator mounted on a 320 CDI diesel engine of a Mercedes-Benz S-Class.,

Experiments with the oilpan were performed in the laboratory so that a wide range of test conditions could be covered. For that purpose the forced cooling by the airstream during driving is simulated by a van. Prior to the experiment we measured the flow velocity of the air at different locations near the powertrain and modeled the local airflow near the cooling fins as function of the speed of the car. This model served as input for the airflow supplied by a van for forced cooling of the cooling fins in our setup. Using such setup we measured an output of the thermoelectric generator of 2 mW ($\Delta T = 20$ K) and 6 mW ($\Delta T = 40$ K), shown in figure 4. The characteristic curve of the generator output is plotted in figure 5. The internal resistance measured at a temperature difference of 45 K decreases by almost 50% compared to measurements with a temperature difference of 15 K.

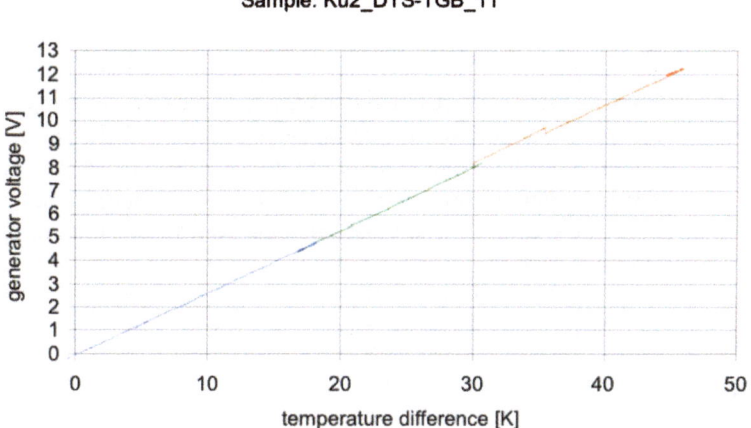

Sample: Ku2_DTS-TGB_11

Fig. 4.   Generator output as function of temperature difference between mounting site on oilpan and temperature measured on the cooling fins of the cold site.

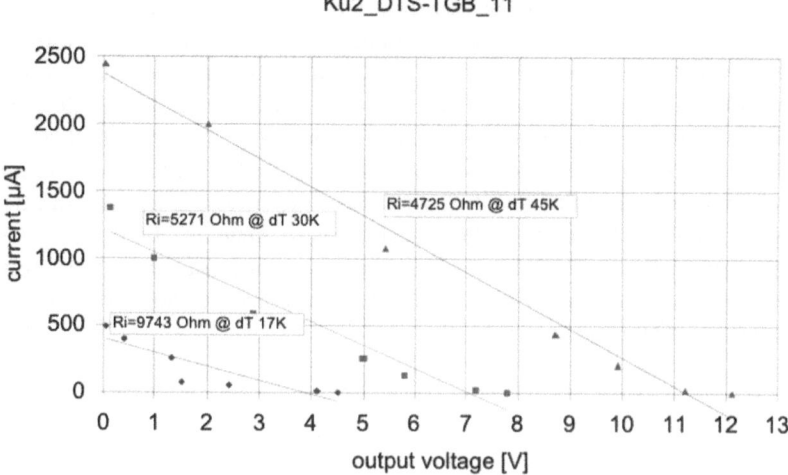

Fig. 5.   Characteristic curve of generator output as function of temperature difference between mounting site on oilpan and temperature measured on the cooling fins of the cold site. The internal resistance measured at a temperature difference of 45 K decreases by almost 50% compared to measurements with a temperature difference of 15 K.

4   Conclusions

We showed that a single chip of the characterized thermoelectric generator is able to supply more than 6 mW at a temperature difference of 45 K. The reliability of this technology still needs improvement. The focus will be on packaging of the chip, especially stacking the foils to built a package. Our experiments furthermore proved the remarkable power density achieved using the described technology. However, the high thermal conductance of each foil layer enables a considerably thermal flow to take place. This behavior is characterized by the comparable low figure of merit of 1% of this particular construction of a thermoelectric generator. A consequence is that a large heat sink is required. The power density per cubic centimeter is thus much lower than for a conventional thermoelectric generator. The overall package dimensions of sensor systems is crucial for automotive applications. As result, a conventionally fabricated thermoelectric generator with state-of-the-art high ZT materials is suggested for future work.

Acknowledgements

We thank DTS, Dr. Stordeur and Dr. I. Stark for their valuable support and service in design and fabrication of thin film based thermoelectric generators matching our special needs.

References

[1] James EP, Tudor MJ, Beeby SP, Harris NR, Glynne-Jones P, Ross JN, White NM: A wireless self-powered micro-system for condition monitoring, Proc. Eurosensors XVI, September 2002.

[2] Qu W, Plotner M, Fischer J: Microfabrication of thermoelectric generators on flexible foil substrates as a power source for autonomous microsystems, Journal of Micromechanics and Microengineering, 11(2):146-152, March 2001.

[3] White NM, Glynne-Jones P, Beeby SP: A novel thick-film piezoelectric micro-generator, Smart Materials and Structures, 10(4):850-852, August 2001.

[4] Hettich G, Vieweger W, Reuss1 H, Mrowka J, Naumann G: Self-supporting power supply for vehicle sensors, 10. Congress "Elektronik im Kraftfahrzeug", Baden-Baden, Germany, 27/28.9.2001, 2001.

[5] Mickle MH, Lovell M, Mats L, Neureuter L, Gorodetsky D: Energy harvesting, profiles, and potential sources, International Journal of Parallel and Distributed Systems & Networks, 4(3):150-160, 2001.

[6] Roundy S, Wright PK, Pister KS: Micro-electrostatic vibration-to-electricity converters, Proceedings of IMECE2002 ASME International Mechanical Engineering Congress & Exposition November 17 22, 2002, New Orleans, Louisiana, November 2002.

[7] Li WJ, Wireless Sensors with Integrated Vibration-induced Power Generator, IROS 2000, 2000.

[8] Stordeur M, Stark I: Thermoelectric Sensor for HF-Power Detector; in: Proccedings of 6th European Workshop on Thermoelectrics, Freiburg, 2001.

A. Bodensohn, R. Falsett
DaimlerChrysler AG
Center of Competence Sensotronic and Microsystems
60528 Frankfurt
Germany

M. Haueis, M. Pulvermüller
DaimlerChrysler AG
Center of Competence Sensotronic and Microsystems
89081 Ulm
Germany

Keywords: autonomous sensor systems, thermoelectric generators, Seebeck

European on Board Diagnostic Catalyst Efficiency Measured on Board and on Test Bench

A. Celasco, F. Cavallino, Powertrain Italy

Abstract

A new European On Board Diagnostic system (EOBD) for real time monitoring of emissions from gasoline engines has been developed, set up and applied in order to satisfy the Euro 3 and Euro 4 regulations. The new system has been required in order to control the performances of fuel injection, closed-loop control and three-way catalyst [1].

The EOBD system has been legislated by the Directives 98/69, 2002/80 and 2003/76 which are mandatory for Euro 3 and Euro 4 configurations. Therefore, the EOBD diagnosis has been developed to check the following systems/components:

▶ Catalyst efficincy.
▶ Oxygen sensor.
▶ Misfiring rate.
▶ Fuel system components.

This article reports preliminary indications to perform a catalyst efficiency monitoring. A computational procedure is applied for evaluating the oxygen amount delivered by the catalyst in order to indirectly estimate the catalyst oxygen storage capacity. The EOBD has been applied with the new catalyst layout (pre and main, close coupled, etc.) in order to meet Euro 3 and Euro 4 Standards (2000/2005).

1 Emission Behaviour in Durability

In order to better characterise catalyst durability performance, a new type of emission representation is now frequently used; it is the so-called "cumulated emissions". Figure 1 shows an example of the three pollutants when driving the MVEGB European cycle (ECE+EUDC). The diagram outlines catalyst behaviour during mileage accumulation in four steps: 0; 15.000; 60.000 and 80.000 km (50.000 miles). This representation gives the advantage of easily representing the importance of ageing linked to cold- and to hot-cycle phases.

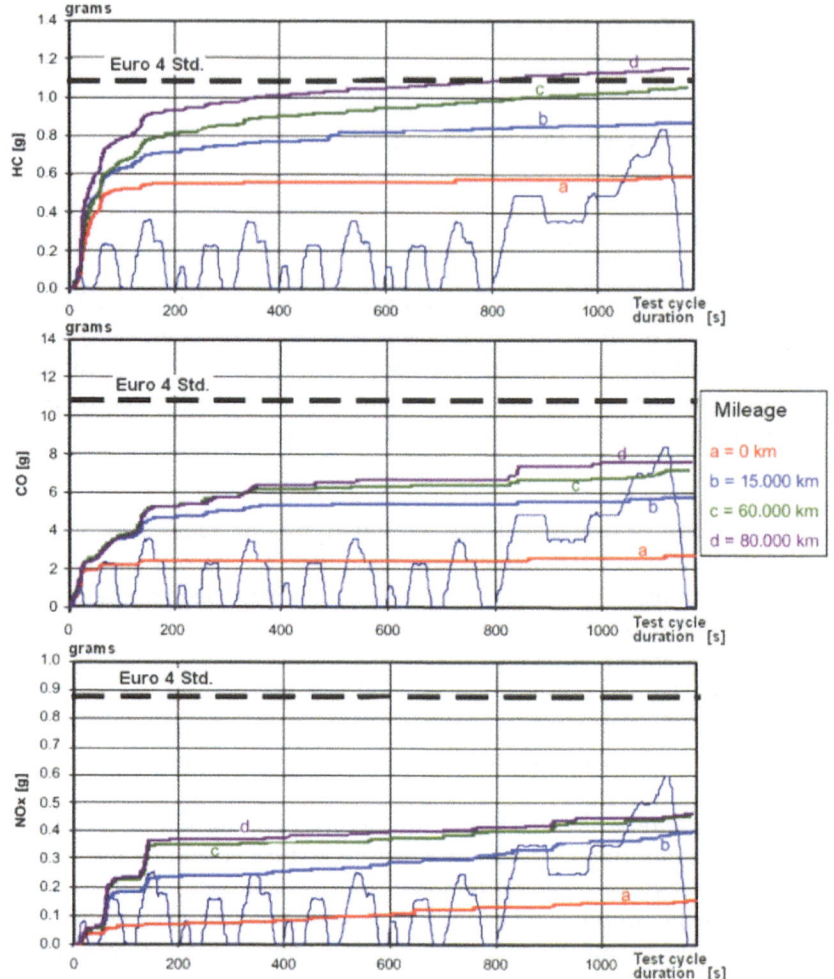

Fig. 1.   Cumulated emissions; Fiat Punto 1,2 Liter Euro 4.

This set of measures refers to a prototype layout: a pre - and main- catalyst system (0.3 and 1- litre) applied to a 1.2 l; 8 valve engine in order to achieve Euro 4 level [2].

It is possible to distinguish the effect of the light-off temperature increase, cold phase, from the factors capable to deteriorate the hot phase: loss of surface area and accumulation of metals capable to poison active wash coat sites.

The main properties that improve light-off are: positioning of the catalyst closer to exhaust manifold, reduction of catalyst/exhaust thermal capacity, proper noble metal selection and cross-frontal area. Some effects that worsen hot-phase performances also impair the cold-phase ones. It is worth mentioning that EOBD monitoring is not effective during the warm-up phase, and its diagnostic capability can take place only during the hot phase.

2   Catalyst Efficiency

After having reached the light-off temperature, catalyst efficiency is very high. Figure 2 shows the emission reduction when A/F control is maintained on the stoichiometric value. Oxidation and reduction reactions are well known and very effective. Unfortunately A/F control is not always so close to stoichiometric as desired; on the rich side it is useful to outline the beneficial effect of water vapour, which promotes the water shift and the steam reforming reactions in order to lower CO and HC. On the lean side, NO_x rise is very steep [3].

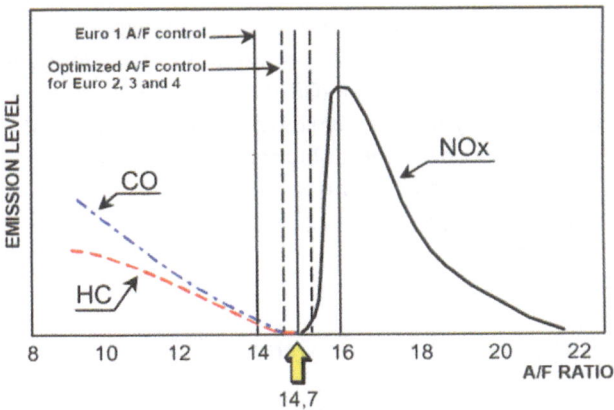

Fig. 2.   Tailpipe emissions versus A/F.

Figure 3 shows the effect of ceria reactions capable of absorbing oxygen in the lean side and to release it in the rich side. In other words, its presence is capable of widening the stoichiometric window. Figure 4 outlines the fact that noble metal and ceria are very well distributed [3]. More specifically figure 4a shows both ceria and precious metal particles with high surface area, while Figure 4b is representative of a catalyst that still presents a sufficiently high surface area, but has lost much of its ceria surface and its interaction with the metals.

Additional reactions (SAE n° 931034)	
Water gas shift $CO + H_2O = CO_2 + H_2$	
Steam reforming $C_xH_y + 2xH_2O = xCO_2 + [2x + (y/2)] H_2$	
"Oxygen Storage" reactions	
Ceria released $CO + 2CeO_2 = CO_2 + Ce_2O_3$	**Ceria stored** $O_2 + 2Ce_2O_3 = 4CeO_2$
RICH SIDE	**LEAN SIDE**

A / F

Fig. 3. Additional reactions.

Therefore, when the catalyst reaches its proper operating temperature, the following key parameters are essential for obtaining the expected performances:

- ▶ Narrow A/F control both in steady state and in transient conditions.
- ▶ Proper Geometric Surface Area (GSA).
- ▶ Noble metal quantity and its distribution on the wash coat.
- ▶ Relative distribution between noble metal and ceria.
- ▶ Relative ceria/noble metal quantity.
- ▶ Proper Oxygen Storage Capacity (OSC).

3   On Board Diagnostic to Monitor Catalyst Efficiency

Catalyst efficiency monitoring is performed by measuring the OSC, which is the only parameter that can be reasonably and continuously monitored on the car, installing a second oxygen sensor downstream of the catalyst. By comparing upstream/downstream oxygen sensor signals, it is possible to obtain an indication of the residual OSC. Dir. 98/69 states that the malfunction indicator must be switched on when HC level, measured on the complete MVEGB cycle, reaches 0.4 g/km. This level is known as the "Threshold" and can be compared with the Type Approval Standard (TAS) equal to 0.2 g/km applicable to Euro 3, which must be fulfilled for 80.000 km.For the time being the Threshold value remains unchanged, even if Euro 4 HC Std. has been reduced to 0.1 g/km for 100.000 km.

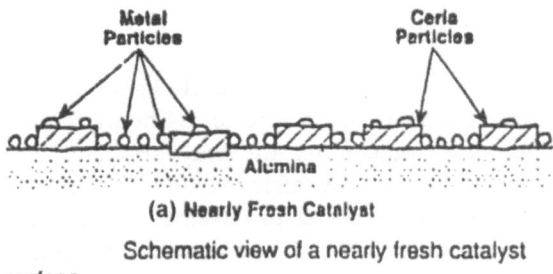

(a) Nearly Fresh Catalyst

Schematic view of a nearly fresh catalyst surface.

(b) Moderately Aged Catalyst

Schematic view of a moderately aged catalyst with high activity, but reduced oxygen storage/release capacity.

Fig. 4.   Active site distribution with fresh and aged catalyst.

OBD catalyst monitoring is performed measuring only HC emissions. It is therefore clear that the upstream and downstream oxygen sensors must be capable of distinguishing between a normal OSC loss of up to 0.2 or to 0.1 g/km, and an unusual loss responsible for emission degradation up to 0.4 g/km. Defining Euro 3 / 4 catalyst is important to guarantee both the stability of the traditional parameters (A/F control, GSA, etc.) as well as the stability of OSC. In this respect the zirconia – ceria mixed oxides have provided the best temperature-stable oxygen storage to date, and are used extensively in close-coupled applications. In fact converter temperatures of 1.000°C are above the phase transition temperature from gamma alumina to alpha alumina. Stabilized alumina is the material that provides the largest surface area on which the active catalytic ingredients have been dispersed for stable operation. When the pores of the catalyst support collapse in the phase transition, they bury the active ingredients and the oxygen storage rare earth materials [4].

Mixed oxides of ceria and zirconium have provided a suitable support surface for the close-coupled catalyst with excellent thermal stability. These temperature-stable oxygen storage components have also provided the basis for the design of oxygen monitors that activate the malfunction indicator light.

4   Catalyst Deterioration

It is wise to underline, figure 5, how the simple heating of the catalyst substrate in an oven with stable air at 1.300 °C for 64 hours is less severe than 80 hours at 1.100 °C, but with substantial A/F ratio variations ($\pm$ 25%). When discussing artificial catalyst ageing for failure simulation, this a key issue if the temperature is to be raised in stable air, or the catalyst is exposed to lower temperatures while determining A/F ratio variations. According to our experiences, this latest procedure appears recommendable and the first one is loosing its importance.

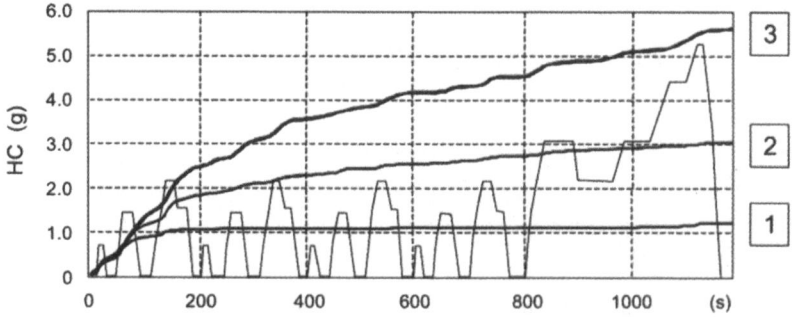

Fig. 5.   Cumulated tailpipe emissions measured with catalyst having different aging.

5   Oxygen Sensor Responses and OSC Determination

The quantity of oxygen stored in a typical catalyst volume (1.5–2.0 litres) ranges between 1.100 and 1.500 mg. When the catalyst is aged, but still satisfies 80.000 km Euro 3 legislated value, HC = 0.2 g/km, OSC ranges between 100 and 300 mg. This value drops below 50 mg when OSC has been seriously reduced and HC level approaches the 0.4 g/km value, therefore determining the switch-on of the malfunction indicator. Figure 6 shows the correlation between Oxygen Storage Capacity and tailpipe emissions.

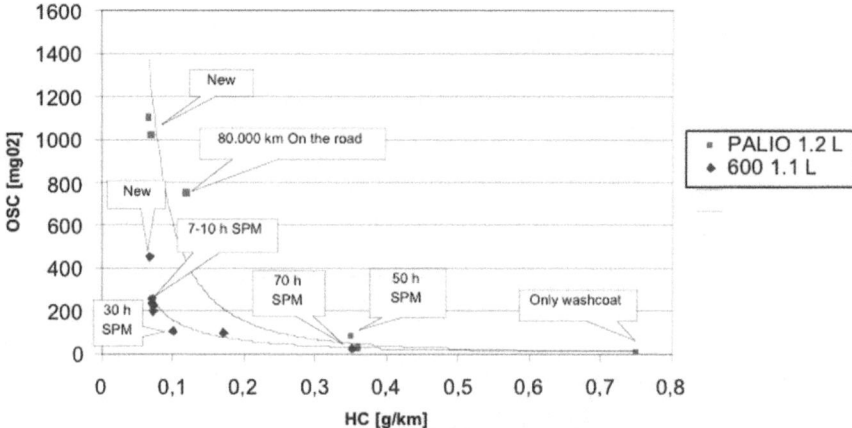

Fig. 6.   Oxygen storage capacity variation versus tailpipe HC measured
driving MVEGB cycle. (SPM: engine test bench at 1.000 °C).

Evaluation of oxygen quantity is carried out by measuring the delay time of
downstream oxygen sensor when a sudden A/F ratio variation is induced
upstream.

Figure 7 shows an A/F step variation, from lean to reach, measured by the two
upstream lambda sensors HEGO (on/off type) and UEGO (linear type); after
approximately 10 s also downstream signal is modified accordingly. In the
upper part of the same figure is shown the behaviour of two UEGO sensors
used as control. It is also outlined the time window determined during the two
transient conditions with the signal B1 − B2. A similar behaviour is recorded
switching from rich to lean. Upper part of the figure shows the behaviour of a
new catalyst, that records a difference, between the two HEGO signals, close
to 2.5 s.

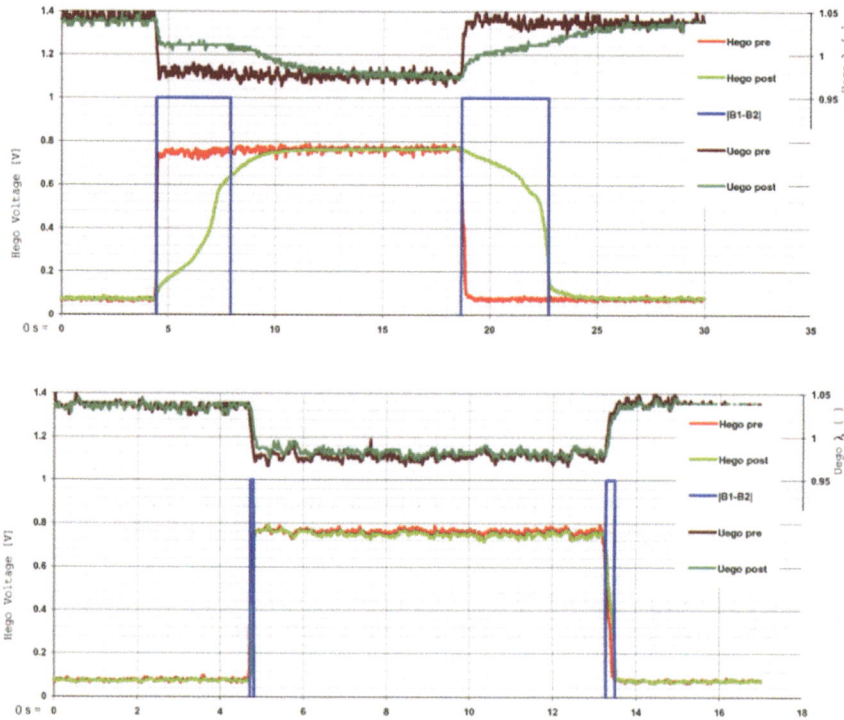

Fig. 7.   OSC for new and aged catalyst (B1 - B2), HEGO and UEGO signals.

The lower part of the same figure reports the behaviour of an aged catalyst; during the A/F transient it is evident that the HEGO signals are the same considering that OSC is close to zero.

Response delays of downstream HEGO compared with the one upstream are very well linked with oxygen storage quantity still available on catalyst surface. It is possible to determinate OSC during A/F variations using formulae specified in (1) and (2):

Oxygen released (mg O_2)

$$OSC_{lean\_to\_rich} = \int_{t\_rich}^{t\_lean} \left[(1 - \lambda_{rif}) \right] xAirFlow\left(\frac{kg}{h} \right) x \frac{1}{3600} x \frac{1000g}{1kg} x \frac{1000mg}{1g} x0.23 \right] dt \quad (1)$$

Oxygen stored (mg O_2)

$$OSC_{rich\_to\_leanh} = \int_{t\_lean}^{t\_rich}\left[(\lambda_{rif} - 1)\right] x AirFlow\left(\frac{kg}{h}\right) x \frac{1}{3600} x \frac{1000g}{1kg} x \frac{1000mg}{1g} x0.23 \right] dt \quad (2)$$

Where:
- Oxygen content in ambient air reaches 23%.
- λ_{rif}: NTK probe reference signal upstream catalys.
- t_{lean} to rich: time in seconds of transient window (the first transient rectangular window).
- t_{rich} to lean: time in seconds of transient window (the second transient rectangular window).

Starting from the above it is possible to define simplified correlation:

O_2 mass released $= 0.23x$ Air Flow/3600 x (1-λ) x $(T_{\lambda,downstream} - T_{\lambda,upstream})$
for $\lambda < 1$

O_2 mass stored $= 0.23x$ Air Flow/3600 x (λ-1) x $(T_{\lambda,downstream} - T_{\lambda,upstream})$
for $l > 1$

Where:
$T_{\lambda,upstream}$ and $T_{\lambda,downstream}$ [s] are the time span to reach 90% of the final value during the rise of signal induced by mixture enrichment and the time span to reach the 95% of the final value during the signal decrease induced by the mixture leaning

Air - flow [kg/h] is the air induced through engine intake
A numerical example of O_2-stored determination is reported [5]:
0.23 x 50/3600 x (1.1-1.0) x 2.5 = 800 mg

Where:
0.23 is O_2 content in ambient air [kg/h]
(1.1 − 1.0) lambda variation
$(T_{\lambda,downstream} - T_{\lambda,upstream}) = 2.5$ s

6 Measurable Parameters on Car

Figure 8 and 9 [5] show a typical response of downstream HEGO (VLAM2) for a new catalyst and for a one definitely aged in comparison with the upstream

HEGO signal (VLAM 1). In particular, the numerical catalyst efficiency is 99% for the new catalyst and 16,94% for the aged one. It is clear that VLAM2 frequency is very low in the first case and high in the second case; in this later case its frequency becomes similar to the one of upstream sensor (VLAM1).

Starting with cold engine, during the first 300 s of the MVEGB cycle, the VLAM2 signal is not active because it does not reach its operating condition.

The useful signal comparison takes place only when the engine reaches its thermal operating condition and the diagnosis becomes active.

At this moment the ECU activates DCASTABT function, which systematically is active to check when proper conditions are reached (catalyst temperature, engine rpm, vehicle speed, etc.) and to calculate catalyst efficiency.

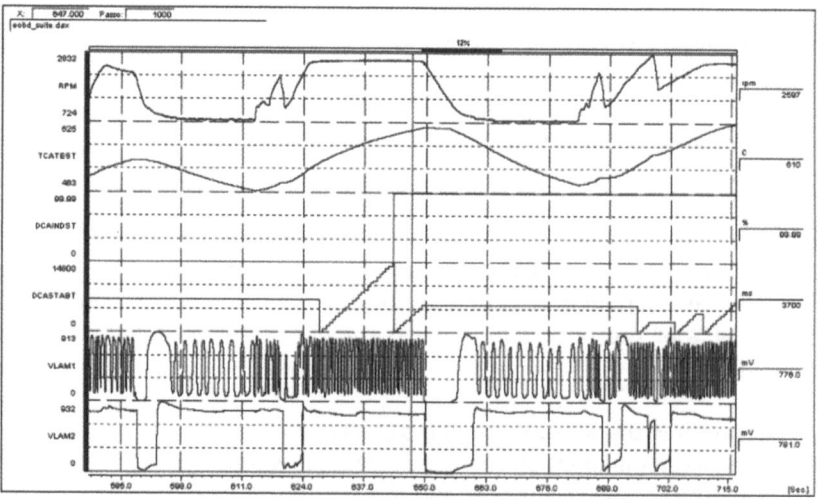

Fig. 8.   Determination of new catalyst efficiency.

In figure 8 catalyst efficiency reaches 99,99%, in figure 9 this value drops to 16,94%.

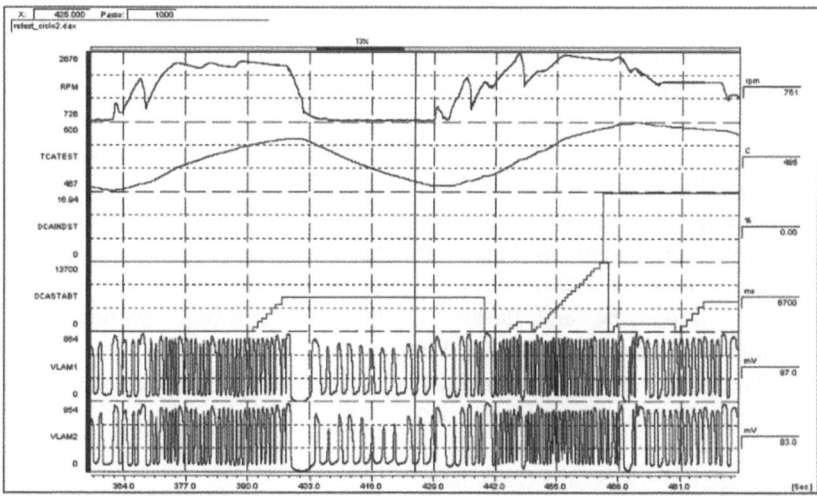

Fig. 9.   Determination of aged catalyst efficiency.

The emission level that activates the malfunction indicator lamp (MIL) corresponds to 0.4 g/km of HC during the MVEGB cycle. This condition takes place when the index of catalyst efficiency, DCAINDST, is about 60%.

7   Conclusions

The new on-board diagnostic system for the analysis of pollutant emissions from gasoline engines has been developed, set up and applied in order to satisfy the Euro 3 and 4 regulations. In the built-up EOBD system - European on Board Diagnostic system - four distinct diagnoses have been developed for detection of faults that are capable of inducing pollutant emission levels, for each kilometre covered by car, exceeding the regulation limit. In this paper the Catalyst Diagnosis has been described and discussed. A computational procedure has been developed for evaluating the oxygen amount stored by the catalyst in order to estimate indirectly the Catalyst Oxygen Storage Capacity. This parameter is the only one capable to detect catalyst efficiency during normal engine operations after initial start-up and thermal transient.

References

[1] "Exhaust emission and fuel consumption of spark ignited engines in the first decade of years 2000", ATA vol. 54 – n°3/4, Cavallino, Celasco, Saroglia, Marzo 2001

[2] "Fuel consumption and emission reduction entering the XXI century", ATA, Boggio, Cavallino, Saroglia, 1999.

[3] "The role of Ceria in automotive exhaust catalyst and OBDII catalyst monitoring", GM/AC Rochester, Fisher, Theis, Casarella, Maham, –SAE 931034, 1993.

[4] "Characterisation of OBDII features of advanced Pd-Rh catalyst and relation to catalyst composition and design", ASEC Manufacturing, Numan, Demison, Williamson, Henk, SAE 980675, 1998.

[5] "Development and set up of an EOBD diagnostic system for low-emission gasoline engines", PhD Thesis, Celasco, 2000.

Aldo Celasco, Francesco Cavallino
F.G.P. Powertrain Italy
Mirafiori Building
Corso Agnelli, 220 Gate 7, Door 206 P.T.
Italy
aldo.celasco@it.fiat-gm-pwt.com

Keywords: EOBD, catalyst monitoring, lambda sensors, emissions

Comfort and HMI

Multifunctional Sensor to Detect Environmental Parameters

N. Pallaro, F. Visintainer, M. Darin, E. Mosca, Centro Ricerche Fiat

Abstract

The SIRPA project aims at developing a multifunction automotive sensor with the integration, on a single CMOS array, of several sensing functions presently provided by different sensors. Such multifunctional sensor will be used to detect a number of critical environmental parameters, such as the luminance of the background and the visibility (in the presence of fog, rain and windshield dimming), while providing additional information about the driving scenario (e.g. curvature of the road, oncoming vehicles, approaching tunnels, ...). The project was launched by Centro Ricerche Fiat and involves Neuricam, ITC-IRST and University of Trento. The sensor is meant to provide some of the required input data for the automatic control of the adaptive headlamp system developed by CRF. In this system the headlamp beams can be modified according to the particular driving and environmental conditions. The prospect of integrating several functions into a single optical CMOS required redesigning existing functions and developing new ones (fog detection). The approach here used was to develop and validate the individual functions separately and afterwards to integrate them onto a CMOS array.

1 Introduction

Nowadays a lot of sensors are used on vehicle or are being developed to detect environmental parameters (luminance, rain, dimming, ...) and the driving scenario (oncoming vehicles, approaching tunnels, lane detection, ...). These sensors provide the necessary input for comfort and safety assistance systems: illumination adapting to the visibility conditions, fog lamp lighting, turning on (off) lamps before tunnel entrance (exit), illumination adapting when crossing other vehicles, automatic activation of windscreen wipers, automatic demisting, automatic speed control, lane warning, etc.

Car makers have to manage this increasing number of sensors which means complexity in terms of sensor housing, cabling, electronic interfaces and actuation strategies. For this reason, component suppliers are working toward the integration of more functions. Yet, at the moment the integration level is limited to a maximum of three different sensors in a single package and the functions involved are non-imaging.

The monitoring of the area in front of the vehicle with one or more cameras allows to collect useful information on the scenario (kind, dimensions and shape of objects and obstacles) to implement comfort and safety functions. Lane warning and lane keeping systems, which are still in progress, are based on the use of a camera (CCD or CMOS array) placed close to the internal rear-view mirror. The format of the camera is often redundant for the function (e.g. VGA area is not completely used in lane warning function).

The former needs (combination of imaging and non-imaging functions, and reduction of number of sensors) can be successfully addressed by a multifunctional integration on CMOS array. The advantages stemming from this solution are: increased reliability, easy-to-use functions control, reduction in number of sensors and components, lower size, easier mounting on vehicle and reduced total costs.

2   Adaptive Headlamp System

2.1 Description of the System

The Adaptive Front-lighting System (AFS) is a frontal illumination system which can "adapt" the lighting beam configuration to the particular driving condition. Within EURIMUS EM34 project [1], Centro Ricerche Fiat developed an adaptive headlamp system characterised by the working modes summarised in the following table.

Mode	Requirements	Features
Town lighting (TL)	To illuminate pedestrian To avoid dazzling	Wide lateral illumination
Country lighting (CL)	To illuminate roadside markers To avoid dazzling	Similar to the standard low-beam, but it illuminates farther and has a wider angle
Motorway lighting (ML)	High speed driving Motorway lanes To avoid dazzling from rear-view mirror	Long range illumination
Bending lighting (BL)	To illuminate the curve in advance	Beam is pointed left/right
Adverse conditions lighting (AL)	To avoid dazzling from wet ground	Wider lateral illumination, less frontal illumination (that generates glare)

Table 1. AFS working modes.

Each mode has its peculiar light distribution (range and divergence) fulfilling the requirements of the correspondent driving scenario. The desired beam configuration is obtained through an optical system, based on micro optics besides the traditional optical elements. The optics lateral movement is based on piezo-actuation, which provides to change the beam configuration in a suitable time interval. In figure1 the Lancia Kappa vehicle demonstrator and some examples of beam configuration are shown.

 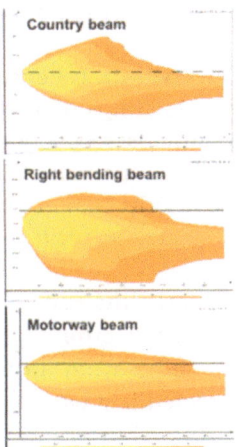

Fig. 1.   Lancia Kappa, demonstrator vehicle equipped with adaptive head-lamp system.

In order to switch automatically from one mode to the other, the adaptive headlamp needs several sensors to recognise the driving condition (vehicle speed, steering wheel angle, ...) and the driving scenario (street geometry, presence of curve, oncoming vehicles, ...), and to measure the environmental parameters (illumination, weather conditions, ...).

2.2 Functional Requirements of the Sensors

This paragraph describes the functional requirements of the main sensors providing data on the driving scenario and the environmental parameters.

The **fog** function measures the density of the external fog to turn the adaptive headlamp beam to the AL mode. The output is the visibility level in metres, within a range of 5 m-150 m, with ±20% confidence. In particularly critical visibility conditions, the system can decide to disable the output of the other optical sensors which could provide wrong data (e.g. curve monitoring).

The **rain** function is aimed at detecting the presence of rain droplets (or snow) on the windshield, in order to actuate the windshield wipers and change their speed according to the rain intensity, and to turn the lighting beam to the AL mode, which minimizes reflections from the wet ground. The sensor should discriminate 5 levels of rain intensity and also phenomena like splash or dew on the external part of the windshield. The device has to work properly also in non-ideal cleaning conditions, due to the wear of the windshield wipers.

The **twilight** function measures the ambient illumination, in order to turn the vehicle lamps on and off automatically in case of scarce or good illumination conditions respectively. Beyond the 'twilight threshold' (approx. 5–10 lux), which is the most important requirement for the adaptive headlamp system, the discrimination of more visibility levels (in the range $1-10^5$ lux) can be used for controlling the vehicle climatic system.

Similarly, the **tunnel** function detects the presence of a tunnel or a very dark passage, in order to automatically switch on and off the car lamps when entering and leaving a tunnel. In particular, the entrance of the tunnel has to be detected in advance, taking the actuation time in account.

Crossing vehicle detection allows to optimise the use of TL and ML modes. Vehicles travelling either in the same or in the opposite direction with respect to the driver are detected, with such an advance as to allow the switching of the beam modes before dazzling the crossing or preceding vehicles.

Finally, the scene in front of the vehicle has to be monitored, in order to detect the presence of a **curve** in advance with respect to the data provided by the steering wheel sensor, and to allow the system to turn to the CL mode.

Noise such as ambient light, flashes, dirt or dimming on the windshield, can influence the function performances. Several working conditions should be therefore taken into account, in order to achieve the required robustness and reliability.

The internal **dimming** of the windshield has also to be monitored (possibly in a predictive way, i.e., with more sensitivity than the human eye) and serves as a warning for the automatic activation and regulation of the air climatic system. This function does not provide direct input to adaptive headlamp system, but it is necessary to enable the above mentioned detection functions.

3 State-of-the-Art

The most automotive environmental sensors integrate more sensing functions (maximum of three different sensors) at package level:
1. rain and light sensors (e.g. by Bosch, Hella, Kostal and Valeo) performing rain, tunnel and twilight detection;
2. illumination sensor (e.g. by First Technology), which can detect the passage through tunnels, the twilight condition and the sun irradiation inside the vehicle.

The optical detectors are mostly photodiodes, so that no imaging function is actually performed. The measured quantity is the collected radiation intensity, which is typically: diffused light from the outside scene for twilight and tunnel functions; direct sunlight for solar functions; scattered/reflected NIR radiation in case of rain sensing. The combination of signals coming from different sensors is then analysed through suited algorithms.

4 Multifunctional Sensor [2]

4.1 Sensor Design Based on CMOS Array Partition

The SIRPA [3] project aims at developing a multifunction automotive sensor with the integration on a single CMOS array of part or all the above mentioned functions, presently performed by different sensors. The approach, claimed by a patent, is to utilise the array as the photo-sensitive element and to divide it into sub-areas, each one being dedicated to one or more functions. The solution proposed by CRF will allow to go far beyond in terms of both number and type of integrated functions, thus reducing system costs significantly.

The acquisition and processing of each area can be optimised by the use of windowing [4] (i.e. reading of sub-areas of the pixel array) and the use of the suitable frame rate and sensor parameters (e.g. gain).

First of all, the concept design was performed, in order to choose the most advantageous sensing 'principle' for each function. The main characteristics of the single functions are reported in the following table.

Function	Active/passive*	Detection principle	Optical Filtering	Orientation
Area monitoring	Passive	Image analysis	Polarizer	Horizontal
Rain	Active	Image analysis	Band-pass	Orthogonal to windshield
Misting	Active	Image analysis/intensity level	Band-pass	Indifferent
Fog	Active	Intensity levels	Band-pass	Horizontal or upward
Twilight	Passive	Intensity levels	None	Upward
Crossing vehicle	Passive	Image analysis/intensity level	Polarizer	Horizontal
(*) Active= with emitter; passive= without emitter				

Table 2. Main characteristics of simple functions.

From the functional specifications and concept design, the resolution and field of view (f.o.v.) were evaluated. These parameters yielded the focal length; thus, the only degree of freedom for the f-number was the lens section. This was not a very strict constraint for active functions- because a narrow aperture of the lens could be partially balanced by increasing the emitter power (within ECE safety norms) - while it represented a critical issue for passive imaging functions, such as the area monitoring. Yet, there was a further specification in our design which actually revealed itself advantageous to solve the problem of integration: it was the viewing orientation. For some functions, like area monitoring, the sensor must "see" in front of the vehicle, while for others (e.g rain) it must be pointed orthogonally to the windshield to have uniform resolution in the image plane. On one side, that set a technical problem of optical axis deflexion, while on the other side it gave a further degree of freedom to the optics sizing.

Another important factor determining the system geometry was the isolation of the optical sub-systems. Indeed, the image planes on the CMOS array are very close to each other (approx. 100 μm separation), so we had to face the problem of noise due to stray light.

The following figure shows an example of CRF proprietary array partition [5,6] in which each sub-area includes both image plane and blind pixels for separation; in particular, it corresponds to the first prototype, which is described in chapter 4.2.

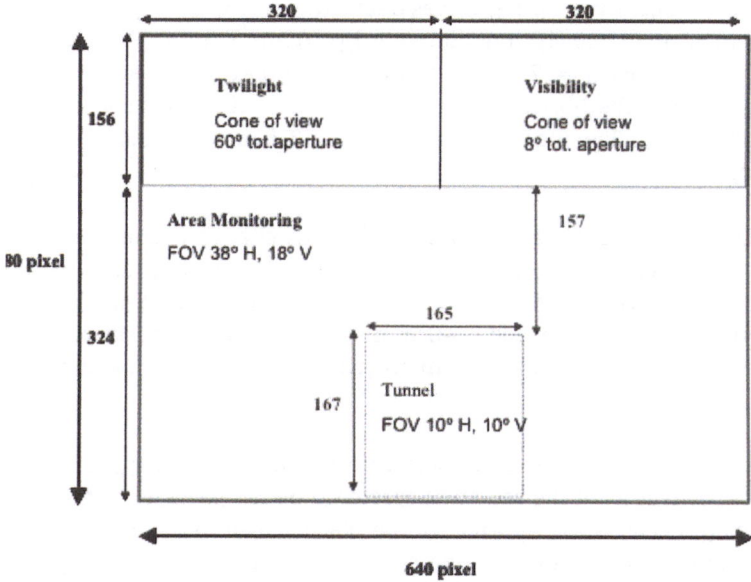

Fig. 2.   VGA CMOS array partition.

4.2 Optical Design of First Prototype [7, 8, 9]

A first prototype of multifunctional sensor was developed (figure 3), integrating onto a single VGA CMOS array the following functions: **visibility, twilight, tunnel** and **frontal area monitoring**.

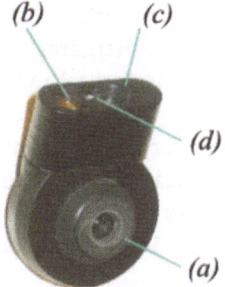

Fig. 3.   First prototype: optical holder to be mounted on VGA CMOS array (45 x 35 x 15 mm); (a) frontal monitoring optics; (b) NIR-LED for visibility function; (c) receiver lens for visibility function; (d) top of the optical fibre for twilight function.

For the **frontal monitoring** optics, it was decided to use wide angle objectives (f = 6 – 8 mm) of small size (10 mm diameter by 15 mm length). With the optimal configuration view (f = 7.5 mm) the objective and CMOS array are oriented about 5° from the horizon to the ground, and, with the f.o.v. and area specified in figure 2, allow for the monitoring of a 3.5 m wide lane of the road from 5 m on.

For the **tunnel** function, a dedicated sub-area was chosen inside the frontal monitoring one, having f.o.v. of 10° horizontally and vertically, and pointing towards the horizon. As described in chapter 2.2, this configuration allows for the recognition of a tunnel at about 40 m distance. Finally, a polarizing filter was placed in front of the optical system, in order to minimize reflections from the vehicle dashboard.

For the **visibility** function, the backscattering technique [10, 11] (emission of NIR beam and collection of the radiation scattered by fog particles) seemed the most promising, because the emitter can be placed close to the detection system, thus maintaining the overall size small. An emitter (NIR-LED) and the receiver collecting lens have respectively the beam divergence and the field of view with partial overlapping. In presence of fog the concentration of small water particles in the overlapping zone produces a backscattering towards the receiver, with an intensity which is proportional to the fog density. Since there are no visibility sensors in the automotive field, a preliminary prototype of fog sensor with a PIN photodiode was developed with good results. Its small size (20 x 20 x 30 mm^2) makes it suitable to be integrated within the internal rearview mirror. For the CMOS based sensor, the collecting lens size was reduced, the emitter power enhanced and the field of view optimised. An optical filter was added for S/N ratio enhancement and a plastic optical fibre (1 mm diam.) was placed after the receiver lens. With the optical fibre carrying the collected signal, the receiver lens (8 mm diameter) could then be placed far enough from the CMOS as to give space to frontal monitoring optics.

Another optical fibre was used for the **twilight** function. The light is collected in a cone of view having 30° semi-aperture and pointing 60° upward. Both fog and twilight fibres are held close to the CMOS dye (400 μm distance), in order to optimise signal intensity (the fibre output beam is indeed divergent).

The problem of optical isolation of the different areas was solved by using an optical window, covered by an absorbing serigraphic coating (mask) on the side facing the CMOS. Such mask, being only 400 μm far from the dye, limits the image planes of the optical subsystems and reduces stray light to tolerable values. In the design of the mask, the manufacturing tolerances of the CMOS package had to be taken into account. For this first prototype, a fine-positioning system was realised to balance the mechanical tolerances which might affect the optics alignment.

4.3 Hardware Architecture [12, 13]

Figure 4 shows the hardware architecture of the adaptive headlamp system which is made of a digital image capturing block (Camera "Eye"), image processing and control unit.

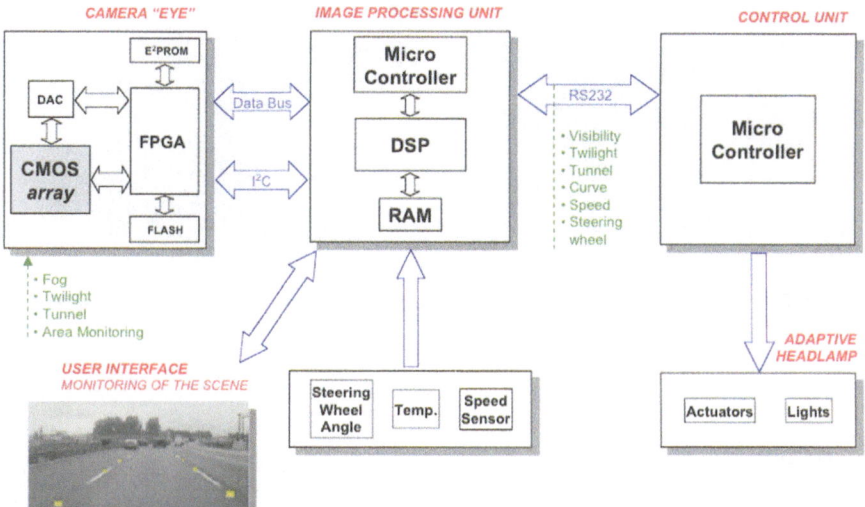

Fig. 4. System architecture of the adaptive headlamp system.

Camera "Eye" is a digital image capturing system specifically designed for automotive and outdoor applications. It is mainly composed of a 0.35 μm CMOS array, an FPGA controller and some peripheral ICs. The sensor, supplied by Neuricam, is a monolithic monochromatic VGA active-pixel which integrates: pixel array, video amplifiers, 10-bit ADC, microprocessor interface and several other support blocks. The response of the pixel is a logarithmic function of the light intensity which allows to cover a high-dynamic-range (120 dB). The FPGA controls the transfer of data from and to the sensor; it is programmed to acquire blocks of pixel (ROIs) and to perform fixed-pattern-noise correction at frame rate. The frame rate is 24 full frames per second.

The image processing unit, which was developed by CRF Engine and Vehicle Division for Lane Warning, is used to acquire and elaborate the digital video stream coming from Eye module. It is composed of a DSP (150 MHz, 24 bit) for the real-time image processing and a microcontroller (50 MHz, 16 bit) to control the I/O signals (digital I/O, CAN B - C, RS232) and the overall system settings (I^2C bus).

Finally, the electronic control unit based on micro controller implements the adaptive headlamp strategies by different outputs (6 actuator drivers, 8 relè, 4 PWM, CAN B – C and RS232). In figure 5 the overall system hardware is shown.

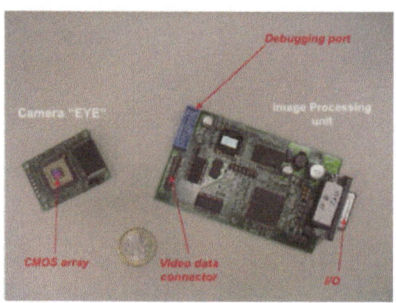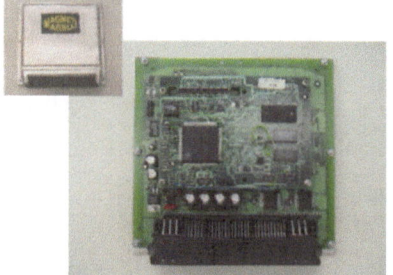

Fig. 5.   System hardware: Eye and image processing unit (left), electronic control unit (right).

5   Results

5.1   Fog Detection

As anticipated in the former paragraphs, a first prototype of visibility sensor for automotive applications, based on a PIN photodiode, has been developed (fig. 6). Visibility range of about 150 m, reduced size, high reliability and low-cost (<15 Euros) are the main features. It is composed of the following components:

▶ Two pulsed NIR LEDs, with a cylindrical lens to shape the emitted beams,
▶ PIN receiver with lens to collect backscattering radiation and an electronic filtering stage,
▶ PIN to verify the cleanness of the windshield in front of the fog sensor,
▶ PIN to check the LED power.

The validation measurements were carried out at the "Laboratoire Regional des Ponts et Chaussees" in Clermont-Ferrand (F), which has a tunnel for artificial and controlled fog generation.

- **LED wavelength**   850 ÷ 880 nm
- **LED beam divergence**   5° x 20°
- **Measurement distance**   5 ÷ 30 cm
- **Visibility range**   10 ÷ 150 m
- **Analog output**   0 ÷ 8 Volt
- **Size**   21x 33 x 24 mm
- **Power supply**   12 V

Fig. 6.   Automotive fog sensor prototype and specifications.

In figure 7 the visibility level (m), measured by a transmissometer, is plotted versus the analogue output (V).

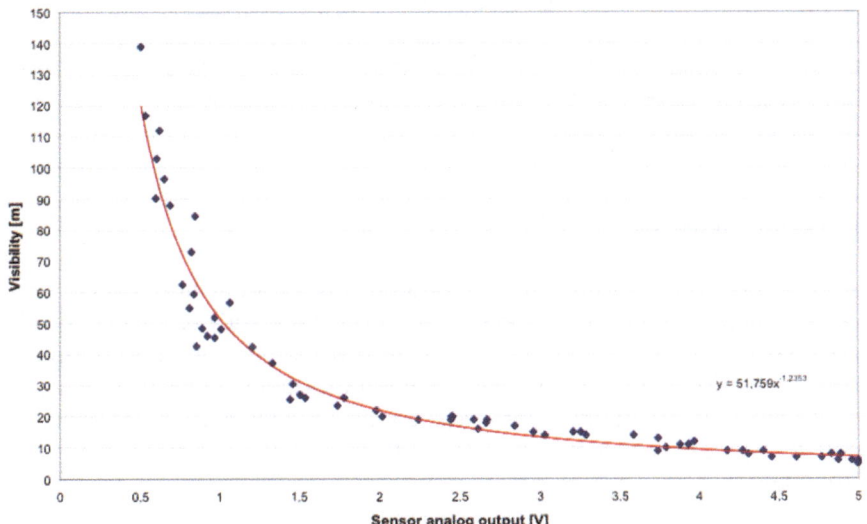

Fig. 7.   Visibility vs. sensor analogue output.

Also for the CMOS array based prototype a set of validation tests have been done in fog-chamber. In figure 8 the visibility level within a range of 5–500 m is plotted versus the average grey scale level (ROI of 55 x 55 pixels).

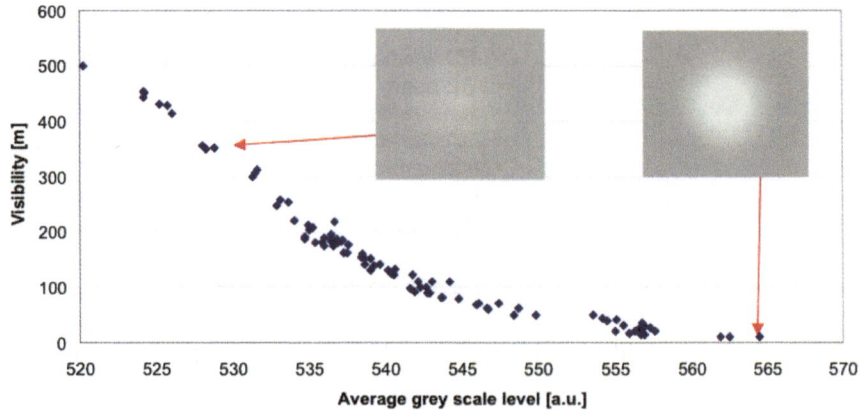

Fig. 8.   Visibility vs average grey scale level.

The intensity of backscattering radiation scales approximately as the inverse of visibility, therefore the output signal is nearly linear, due to the logarithmic response of the CMOS pixels.

5.2 Twilight Detection

Figure 9 shows the transfer function between the external illuminance, measured by a luxmeter, and the average grey scale level (ROI of 55 x 55 pixels). The CMOS logarithmic response allows a good resolution over a wide range of lighting conditions without the need of iris adjustment. In order to discriminate different environmental scenarios, specific ranges must be chosen: twilight (0–10 lux), overcast sky (10–1000 lux) and indirect solar light (> 1000 lux).

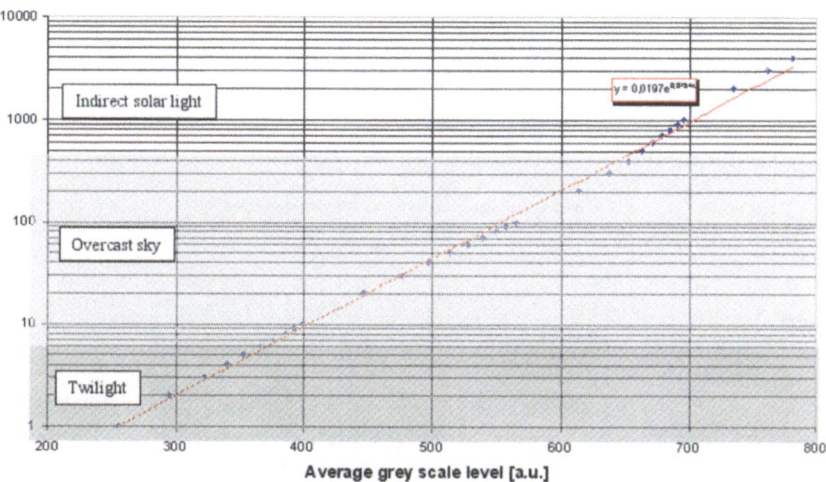

Fig. 9.   Illuminance vs average grey scale level.

5.3 Tunnel Recognition

The aim of this function is to recognize the approach and the passage through a tunnel and to discriminate it from a bridge. The simple and fast algorithm is based on the comparison between two different ROIs contained within the tunnel function subarea, as shown in figure 2. Several tests were carried out with the vehicle demonstrator in order to develop the algorithms and to improve the function robustness. In particular, different roads (urban and extra-urban, motorways) and light conditions were considered.

In figure 10 the algorithm response versus the acquired frame during a typical approach and transition into a tunnel is shown and compared with the output of a commercial twilight/tunnel sensor. The developed function is able to recognize in advance the tunnel entry (and exit) and this is very important when the vehicle speed is high on motorway.

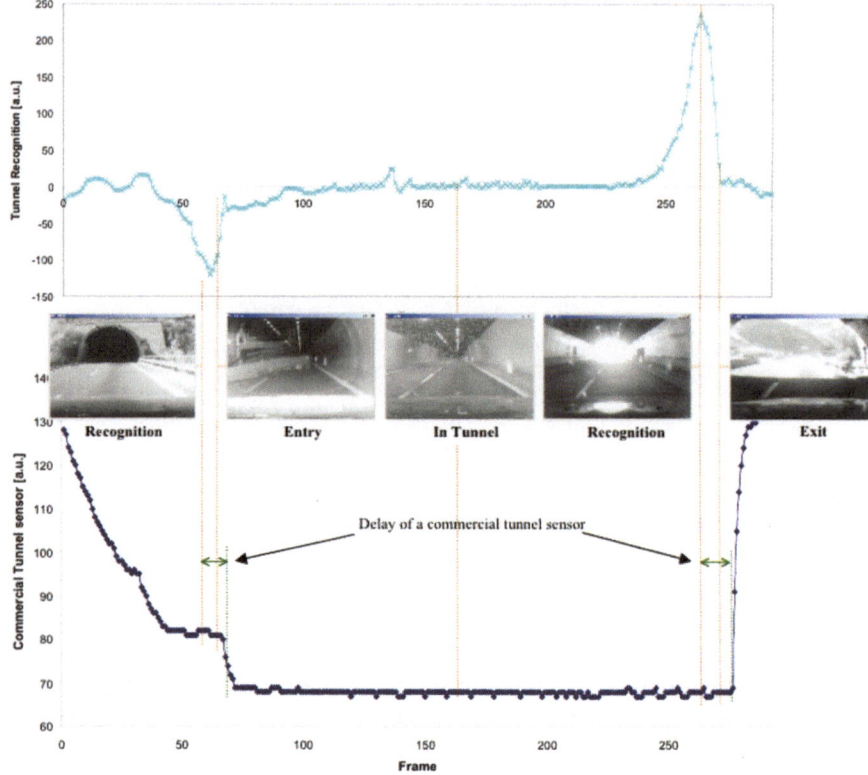

Fig. 10. Example of tunnel analysis.

In the test site above, the car runs at 83 km/h and the algorithm recognises the tunnel entry and exit respectively 35 m and 46 m before.

Another important characteristic is the capability of discriminate the tunnel from a bridge. In the next figure the response of our sensor is compared with a commercial tunnel/twilight sensor. There is not the negative peak, characteristic of tunnel recognition, and the positive peek, associated to the exit, is lower than the tunnel exit peak.

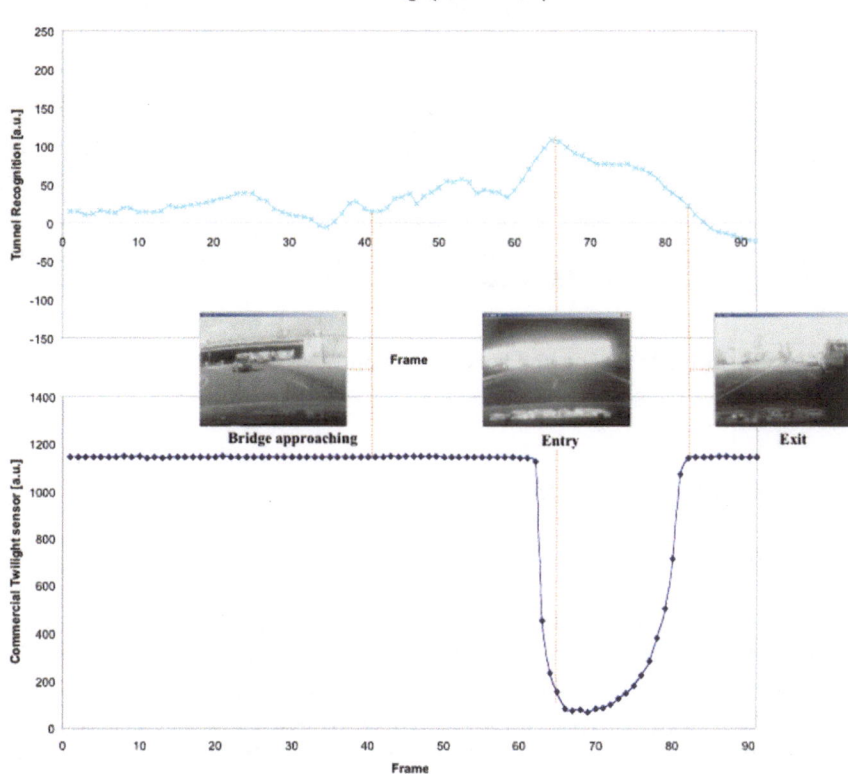

Fig. 11. Example of bridge analysis.

5.4   Curve Monitoring

The developed function belongs to the class of segmentation problems, whose main goal is to divide an image into parts having a strong correlation with objects or areas contained in the image. Therefore, the core of the algorithm is based on methods of thresholding, edge and region segmentation. In the specific class of edge-based image segmentation, the Hough transform segmentation was used to detect objects of known shape (lane) within the image.

In figure 12 an example of road scene with the detected lanes are shown. The angle between lane and vehicle directions and the distance of vehicle from each lane (left and right) are evaluated from the acquired and processed images. Then, the outputs are sent to the control unit of the adaptive headlamp system to bend the lighting beams in advance with respect to the information provided by the steering wheel sensor.

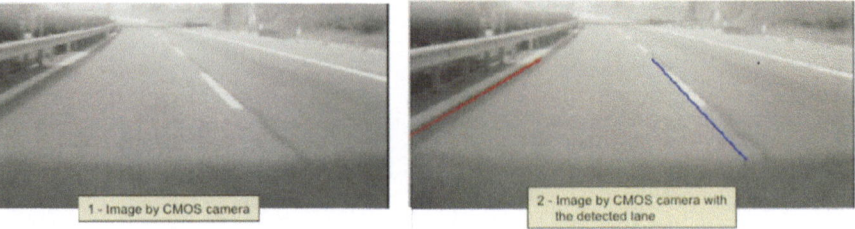

Fig. 12. Example of lane and curve detection.

6   Developments in Progress

6.1   Optical Solutions

Presently, a novel optical solution is being studied, considering non-standard optical elements, such as micro-lenses arrays. The goal is to increase the integration level by replacing discrete optical components with structures constituting of more components (lens, prisms) into the same plastic element. A first solution has already been designed: it is a hybrid system, composed of both lens arrays and traditional lenses, having the suited characteristics to implement the former functions (visibility, twilight, tunnel and curve monitoring) and to add new ones (rain, misting and crossing vehicles). Actually, the algorithms performing the rain function are still in progress, as explained in the next paragraph.

6.2   New Functions to be Integrated

The task of recognizing rain and misting conditions can be attained through the illumination of a limited area of the vehicle windscreen by an NIR source. Without rain drops on the windshield, no NIR light of the emitters reaches the sensor. On the contrary rain drops act as curved mirrors, reflecting the source image in direction of the collecting optics and are thus revealed by edge-detection algorithms. In case of misting, a Lambertian scattering takes place and the source appears like a halo, with different intensity – more or less bright – and position in case of internal or external misting [14, 15].

Another function which is being developed is the crossing vehicle detection. The main requirement is to detect the vehicle with a considerable advance, in order to allow the adaptive headlamp system to change lighting mode before the dazzling happens. The fact that we are dealing with the recognition of car lamps is advantageous, because the light spot appears larger than the size of the illumination optics (which is approximately 15 x 15 mm^2 for headlamps). The dimension of the required area are approximately 320 x 240 pixels.

7 Conclusions

A multifunction automotive sensor with the integration on a single CMOS array of several sensing functions, presently provided by different sensor, was developed within SIRPA project. It is finalised to detect several environmental parameters and to provide additional information on the driving scenario. These output are needed for the automatic control of the adaptive headlamp system, developed in CRF, in which the lighting beams can be modified consequently.

The prospect of integrating several functions into a single optical CMOS required redesigning existing functions and developing new ones (as is the case for fog detection). The approach was to develop and validate the individual functions separately and to integrate them onto a CMOS array.

The concept of multifunction sensor in terms of CMOS array partition and optical solutions has been described. A new automotive fog sensor based on low-cost components (PIN and NIR-LED) and stand-alone electronics was developed to validate the working principle. Then visibility was integrated onto a multifunction sensor together with the curve monitoring and twilight/tunnel functions.

The system architecture is composed of the following modules: VGA logarithmic CMOS array, FPGA based camera interface, embedded image acquisition and processing unit based on DSP and micro-controller, a control unit with the actuation strategies of the adaptive headlamp system.

The main results of each sensing function have been shown and discussed.The project is in progress and further improvements are expected: novel optical solutions, considering non-standard optical elements, such as micro-lenses arrays to increase the integration level and reduce manufacturing costs; implementation of three other functions such as rain, dimming and crossing vehicles detection.

References

[1] EURIMUS (Eureka Industrial Initiative for Microsystem Uses), EM34, Micro-optics/micro-mechanics for adaptive lighting systems,partners: Automotive Lighting (Italy), Atmel (France), Cedrat Technologies (France), Centro Ricerche Fiat (Italy), M. Marelli (Italy) .

[2] N. Pallaro, F. Visintainer, M. Darin, E. Balocco, E. Borello, M. Gottardi, N. Massaro, Multifunctional sensor to detect environmental parameters, AISEM 2003, Trento, 11 February 2003.

[3] SIRPA (Sviluppo di Sensori Integrati per il Rilievo di Parametri Ambientali), project funded by the Autonomous Province of Trento (Italy), partners: Centro Ricerche Fiat (Italy), ITC-IRST (Italy), University of Trento (Italy), Neuricam (Italy).

[4] Control circuit for image array sensors and automatic headlamp control. Inventors: J. Bechtel, J. Stam; applicant: Gentex Corporation; international patent WO 99/14943, publ. 25 March 1999.

[5] Sistema di visione integrato multifunzionale con matrice in tecnologia CMOS o CCD. Inventors: N. Pallaro, F. Visintainer, P. Repetto, E. Borello, B. Pairetti; applicant: Centro Ricerche Fiat S.c.p.a.; Italian patent TO2002A000950, filed 5 November 2002.

[6] Dispositivo di rilevamento installabile lungo una strada per il rilevamento di condizioni ambientali ed il monitoraggio e la sorveglianza del traffico. Inventors: N. Pallaro, F. Visintainer, P. Repetto, M. Darin, E. Mosca, L. Liotti; applicant: Centro Ricerche Fiat S.c.p.a.; Italian patent TO2003, filed 2 October 2003.

[7] E. Hecht, Optics, 4th ed., Addison Wesley, 2002.

[1] R.E. Fischer / B. Tadic - Galeb, Optical System Design, SPIE Press, Mc Graw – Hill, 2000.

[8] F. Grum, R.J. Becherer, Optical radiation measurements, Vol.1, Academic Press, 1979.

[9] H. C. Van de Hulst, Light scattering by small particles, Dover Publications, N.Y., 1957.

[10] J.D. Crosby, Visibility Sensor Accuracy: what's realistic?, presented at the 12th Symposium on Meteorological Observations and Instrumentation 2003 American Meteorological Society Annual Meeting, Long Beach, CA, 9-13 February 2003.

[11] M. Sonka, V. Hlavac, R. Boyle, Image Processing, Analysis, and Machine Vision, 2nd Edition, ITP, 1999.

[12] R.C. Gonzalez, P. Wintz, Digital Image Processing, 2nd Edition, Addison Wesley, 1987.

[13] Moisture sensor and windshield fog detector. Inventors: J. Bechtel, J. Stam, K. Roberts; applicant: Gentex Corporation; U.S. patent US6097024/00, publ. 1 August 2000.

[14] Rain sensor using statistical analysis, Inventors: J. Tenenbaum, P.A. Hochstein; applicant: Valeo Electrical Systems; U.S. patent US6144022/00, publ. 7 November 2000.

Nereo Pallaro, Filippo Visintainer, Marco Darin, Emilio Mosca
Centro Ricerche Fiat
Trento Branch, MicroSystem Group
Via dei Solteri, 38
38100 Trento
Italy
nereo.pallaro@crf.it

Keywords: multifunctional sensor, array partition, CMOS imager, image processing
camera, environmental parameters, visibility sensor, curve recognition,
adaptive headlamp system

Windshield Fogging Prevention by Means of Mean Radiant Temperature Sensor

S. Mola, G. Lo Presti, M. Magini, N. Presutti, C. Malvicino, Centro Ricerche FIAT

G. Bisceglia, M. Mandrile, A. Tarzia, G. Caviasso, Fiat Auto

Abstract

To improve driving safety it is of outmost importance to guarantee that the windshield never starts fogging by means of an automatic and reliable procedure. To reach this goal and keeping always in mind comfort and fuel consumption, a fogging condition estimation algorithm has been developed. It requires a Mean Radiant Temperature (MRT) sensor and a humidity sensor. The MRT sensor is placed on the cabin compartment roof and points towards the dashboard and the windshield. Therefore it may be used to estimate windshield inner surface temperature. Combining this last piece of information with cabin relative humidity measured by the humidity sensor, the estimation algorithm forecasts fogging conditions so that the automatic thermal control system is able to take actions before fogging starts to appear. Centro Ricerche Fiat has already developed for Fiat Auto and patented a cabin thermal control that uses as feedback just MRT and blown air temperature one. This approach is not only safer, but it allows also cost savings with respect to state of the art sensor layout.

1   Introduction

One of the most important tasks that the air conditioning system has to perform on a vehicle is to assure perfect visibility by removing moisture condensed on the inner windshield surface. This is obtained by blowing warm air with an appropriate velocity and distribution through the defrost outlet. Usually, even if there is an automatic climate control unit, the driver is required to interact with it to start defog or defrost procedure by pressing a button and/or turning on the compressor.

This has two consequences on safety:

▶ The action takes place when moisture formation is already affecting driver visibility.
▶ The driver is required to perform an action that is not related to driving, thus affecting his/her level of attention.

Moreover, this may also have an impact on thermal comfort, since defrost/defog procedure usually sets airflow to maximum level and temperature blending to maximum heat. It may not be possible to assure both defogging procedure and thermal comfort, since the former is obviously a priority. But there are several other strategies that may benefit from reliable estimation of incipient fogging conditions. In particular, those that require control of the recirculation flap, such as air quality systems based on gas sensor, or energy saving strategies aimed at reducing thermal load on heater and evaporator.

Impact on fuel consumption is also worth to be mentioned. One of the devices that can help in defogging is turning on the compressor, provided that the external temperature is not too low (usually in the climate control unit there is a constraint regarding it). This results in condensation of the water vapour contained in the air to be treated by the heater, allowing to send drier air to the windshield, but it increases also fuel consumption: therefore, to know in advance that fogging condition is bound to appear may avoid the need for the compressor to be turned on.

From the above considerations, it is clear that to know when incipient fogging occurs can be very helpful for several reasons: safety, comfort and fuel consumption.

Generally speaking, there are two ways to get this piece of information: to detect it by means of sensors that are directly influenced by the condensation itself, or to estimate it by means of models or via sensors that measure physical quantities that influence the phenomenon. The different approaches are compared in [1]. If the approach requires additional sensors, in the analysis of the solution it is very important to evaluate if they can be useful also for other functions of the air conditioning system, or if the offer potential for additional functions (as an example, humidity control), so to perform the correct cost benefit evaluation.

Centro Ricerche FIAT has chosen a sensor set up that include a Mean Radiant Temperature sensor and a humidity sensor. This solution allows for incipient fogging condition detection and is perfectly integrated with other functions of the air conditioning system: thermal comfort and humidity control.

2   Technical Background and Sensor Layout Description

Since 1998 Centro Ricerche FIAT has been using Mean Radiant Temperature (MRT) sensors for a better estimation of comfort conditions in car cabin compartments. MRT performs an integral measurement of radiant temperature of all the surfaces over a determined field of view. A methodology for identifying perceived thermal comfort estimators by means of MRT has been developed for Fiat Auto and successfully applied in several prototypes [2]. The approach (patented by Centro Ricerche FIAT) requires the MRT to be placed in the middle of the roof, pointing towards the windshield and the dashboard, as shown in figure 1.

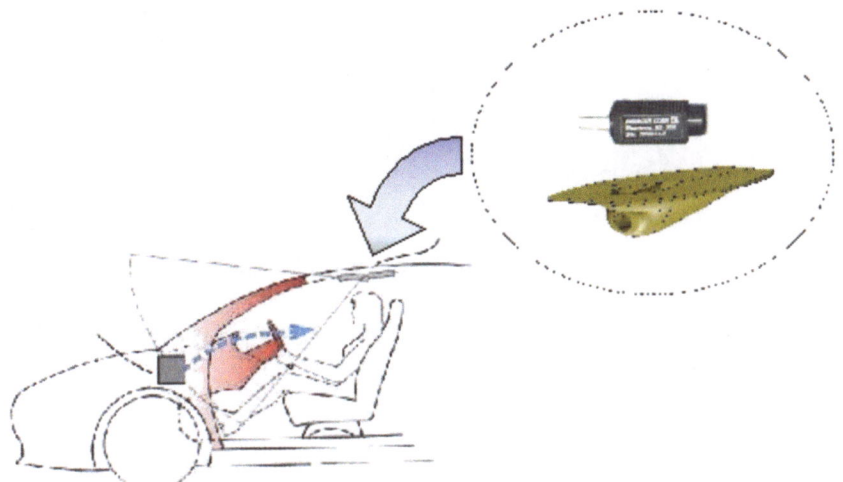

Fig. 1.   Position and field of view of MRT.

It has been shown that this approach allows both sensors number reduction (no need for car cabin compartment air temperature sensor and for sun radiation sensor) and improvement in perceived comfort estimation. Since MRT points towards the windshield Centro Ricerche FIAT has started a research activity so to evaluate the possibility of using information coming from MRT (in combination with a humidity sensor) in order to estimate incipient fogging conditions.

Therefore, since a humidity measurement is necessary, a capacitive humidity sensor has been added near the MRT one (see figure 2):

Fig. 2.   MRT sensor and humidity sensor placed in the car.

Summing up, the A/C control unit will use the signals coming from the following sensors:

▶ Mean Radiant Temperature.
▶ Relative Humidity.
▶ External Air Temperature.
▶ Blown Air Temperature.

An algorithm has then been developed so to be able to forecast incipient windshield fogging conditions. The block diagram of the proposed approach is the following:

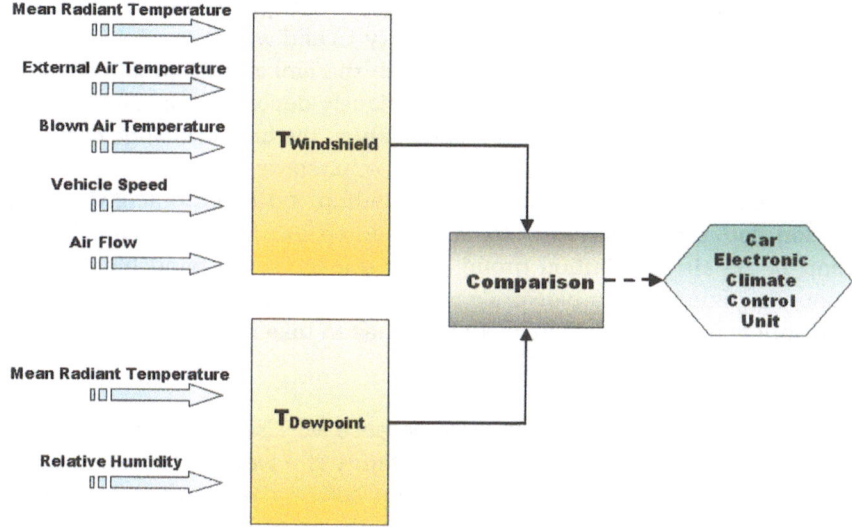

Fig. 3. Algorithm to be implemented in ACU.

3   Algorithm Description

From a theoretical point of view there is the need to estimate or to measure dew point temperature of cabin compartment air and windshield inner surface temperature, and then to make a comparison between them. To measure windshield inner surface temperature would require a dedicated sensor, that could not be used for other purposes. Centro Ricerche FIAT has therefore chosen to estimate it by means of a simple, lumped parameters model:

$$C\frac{dT_w}{dt} = G_i(T_{exch} - T_w) + G_e(T_e - T_w) \tag{1}$$

The quantities indicated in (1) are defined here below:

T_w Windshield inner surface temperature [°C]
T_{exch} Inside air exchange temperature [°C]
T_e Ambient air temperature [°C]
C Windshield heat capacity [J/°C]
G_i Windshield heat exchange coefficient, inner (cabin) side [W/°C]
G_e Windshield heat exchange coefficient, outer side [W/°C]

In this modelling approach, the windshield is just a "point" where all the glass mass is concentrated (with the heat capacity C) and whose temperature is T_w. This "point" exchanges thermal power with the ambient air proportionally to heat exchange coefficient G_e, which is obviously dependent on car velocity. It exchanges also thermal power, proportionally to the heat exchange coefficient G_i, with what we have called "Inside air exchange temperature". The reason for this definition lies in the fact that, depending on the position of the air distribution flap, the temperature to be considered for the heat exchange is different. If the air distribution group sends the air to the windshield (defrost mode), T_{exch} must be the blown air temperature. When the air is blown from other outlets (i.e. floor or vent mode) one has to take into account cabin compartment air.

This model approach is necessary also because MRT sensor does not measure T_w. As it is shown in [2], MRT is representative of a weighted mean of quantities related to comfort. It takes into account both surface temperature and air temperature. Therefore, it can be used for:

▶ Dew point temperature, in combination with relative humidity measurement.
▶ T_{exch}, when not in defrost mode.

It is easy to understand that this is a simplified picture of a very complex physical phenomenon. But we have to take into account that the final goal is to set up an algorithm that can run on a standard production control unit, therefore we have to keep it as simple as possible, and at the same time it has to reliable.

Therefore, heat exchange coefficients cannot be calculated from literature formulae, but have to be identified thanks to an experimental methodology, keeping in mind that we have to simulate as T_w the temperature of the part of the windshield where it is most likely that condensation takes place. From [1], this location is behind rear mirror (this has been confirmed also by testing performed at Centro Ricerche FIAT).

From the above discussion, it is clear that this model approach must be coupled to a testing methodology and to an identification procedure so to find out the value of G_i and G_e, which are the unknown quantities that cannot be estimated a priori.

4   Testing Methodology and Identification Procedure

The car has been equipped with the following sensors:

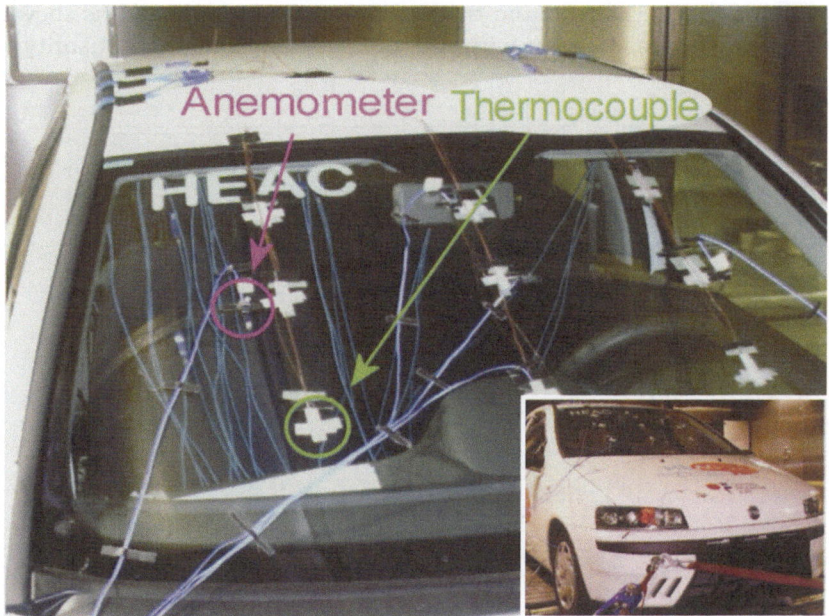

Fig. 4.   Car experimental set up.

The first series of testing has been performed with the following goal: to identify G_i end G_e so to be able to reproduce, by means of the model, the time behaviour of T_w, as measured by the sensor placed behind the rear mirror.

The car has therefore been placed in a climatic chamber where humidity and temperature can be controlled. The testing has been performed in the following way:

▶ Ambient conditions (temperature and relative humidity) are fixed and kept constant throughout the testing.
▶ Engine is turned on.
▶ A value of air flow is selected.
▶ Acting on air distribution flap, warm air is sent to the windshield (warm up period).
▶ After the warm up period, acting again on air distribution flap, the air is blown through the front outlet (cool-down period).

When the cool-down period is over, another value of airflow is selected and the procedure is repeated until all airflow values of interest have been tested. Then, other values of external velocity and ambient conditions are imposed. The methodology has been optimised so that no more than two weeks of testing are necessary. The resulting time behaviour for measured T_w is shown in the following figure, together with the corresponding simulated quantity:

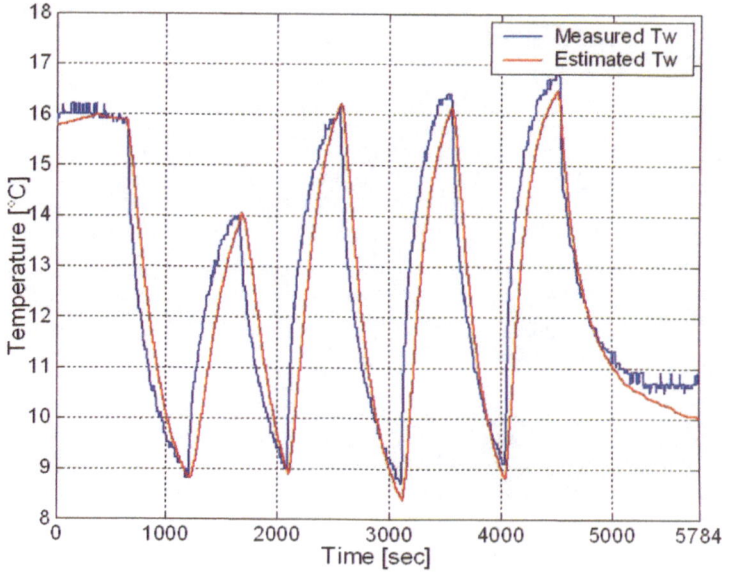

Fig. 5.   Testing and simulation comparison.

The simulated curve has been obtained by means of equation (1). An algorithm developed in Matlab-Simulink that minimizes the error between acquired T_w and measured one has identified values of G_i end G_e. Obviously, some of the tests are used for identification and some for validation.

The above results show that T_w cab be reproduced by means of model (1) in a reliable way. The other quantity that it is necessary to reproduce is the dew point temperature. This quantity is not simulated, but calculated from MRT and relative humidity sensor signals. Therefore, a series of validation tests for the whole approach have been performed. The comparison to be made is between the incipient fogging condition measured by the optic sensor and the corresponding value calculated by the algorithm, as shown below:

The test has been performed on road under real operating conditions. It is easy to see that every fogging condition detected by the optic sensor is also detected by the estimation algorithm. Moreover, the estimation algorithm is always predictive with respect to the optic sensor, as expected. This allows to make the strategy more effective from an energetic point of view, since there is no latent heat to provide in order to prevent fogging.

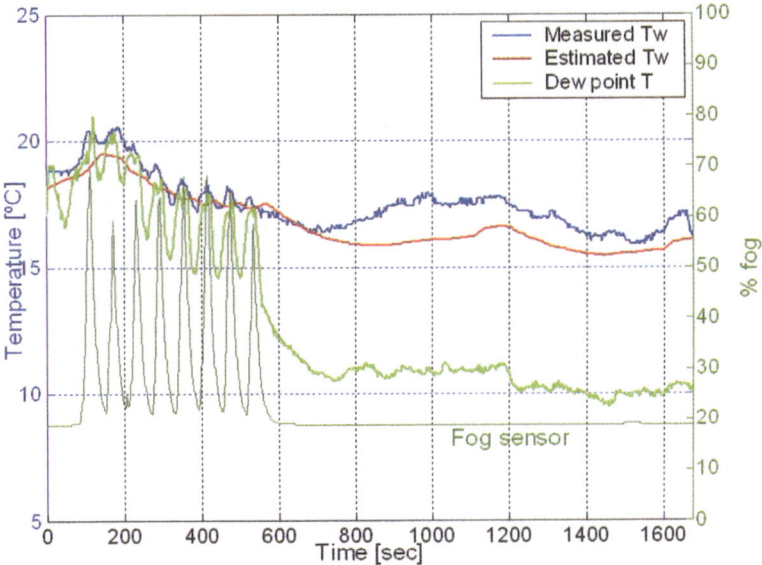

Fig. 6. Validation for incipient fogging detection.

5 Potential for Integration with Other Climate Control Features

As mentioned above, from cost benefit analysis point of view, it is very important to evaluate the potential for other functions. We have already mentioned that MRT sensor can be cost effectively used as feedback for the climate comfort. MRT sensor combined with an externally controlled variable displacement compressor is an important Fiat development line in order to reduce the impact on fuel consumption increase given by the air conditioning system [3].

In [3] it has been shown that thanks to control and by using this component, fuel consumption increase due to air conditioning system can be reduced up to 50%. This comes from the fact that in most cases we can avoid reheat, just transferring to the air the right amount of refrigerant power that is needed to assure comfort. This implies more or less to have a control on the evaporation pressure level.

But if we can control this last quantity, this implies that we can vary (within a certain range) also the humidity content of the blown air. Obviously, we cannot humidify: but we can modulate the dehumidification of the air that it is treated by the evaporator. This offers a potential for cabin air humidity management (it is clear that we cannot guarantee in all conditions that minimum impact on fuel consumption and desired humidity level are satisfied at the same time).

With the sensor layout that we have described in this paper, this function can be assured without the need for additional sensors (obviously, provided that we have an externally controlled variable displacement compressor). This feature has been implemented on a prototype, allowing the user to select both comfort temperature and humidity level. Since impact on consumption may be un important issue for mobile air conditioning systems in the near future, it is likely that externally controlled variable displacement compressor may diffuse. Therefore, humidity management might also diffuse.

6   Conclusion

We have shown that adding a relative humidity sensor to a MRT one, we have a reliable estimation of incipient windshield fogging condition. It is a predictive method: we get the information when condensation has already taken place, thus requiring more energy to remove it (because of phase change). It is not sensitive to dirt or to smoke. Moreover, there are already available on the market MRT sensors integrated with relative humidity ones. The calculation and testing necessary to implement this approach on a new car requires no more than two weeks.

The proposed sensors layout provides signals both for climate control and for humidity management. With respect to standard layout, it allows a reduction in the number of sensors. Standard production FIAT Auto approach to guarantee the same functions (thermal comfort and incipient fogging estimation) requires 4 sensors: cabin air temperature, ambient temperature, solar sensor, optic fogging sensor. The one described here, 3 sensors: ambient air, MRT sensor, humidity sensor. It has to be taken into account that optic fogging sensor, besides being sensible to dirt, it is also quite costly. If we add also humidity management function, standard production layout would require an additional relative humidity sensor. As discussed above, this approach does not require additional sensors for humidity management. In this case, there would be two sensors less.This approach, thanks to its high level of integration, offers high potential for cost savings.

References

[1] Thomas M. Urbank, Sean M. Kelly, Timothy O. King, Charles A. Archibald, "Development and Application of an Integrated Dew Point and Glass Temperature Sensor", SAE 2001-01-0585

[2] S. Mola, M. Magini, C.Malvicino, "Measuring Thermal Comfort in Vehicles: an approach using mean radiant temperature", XI Sensors and their application, London, September 2001

[3] S. Martini, M. Magini, S. Mola, G. Lo Presti, C. Malvicino, G. Caviasso, A. Tarzia, J. Benouali, D. Clodic, "High efficiency air conditioning: improving comfort and reducing impact on fuel consumption", VTMS 6, Brighton, May 2003

Stefano Mola, Giulio Lo Presti, Massimo Magini, Nicola Presutti, Carloandrea Malvicino
Centro Ricerche FIAT
Strada Torino 50
10043 Orbassano (TO)
Italy
stefano.mola@crf.it

Gennaro Bisceglia, Massimiliano Mandrile, Antonio Tarzia, Guglielmo Caviasso
FIAT Auto
Corso Agnelli 200
Torino
Italy

Keywords: Fogging Prevention, Climate Control, Mean Radiant Temperature Sensor, Sensors Number Reduction.

Networked Vehicle

Sensor Integration in the Mechatronic Environment

G. Teepe, Motorola

Abstract

Next generation car architecture will be marked by seamless component integration. The „Mechatronics Concept" enables composable architectures in the car and will lead to increased development quality. The upcoming of LIN™ and FlexRay™ based networking systems in the car is in line with this new development paradigm, marked by model based system development, where the system design is supported by component modelling at each stage. This will lead to results which are „correct-by-construction" and will shorten the qualification cycle times concurrently with a significant increase in quality and reliability. Sensor design will have to comply with the new integration paradigms into the car electronic system. This will ultimately result into a software centric car architecture, which is being defined by the newly established AutOsar consortium. It will ultimately resolve the lifecycle disconnects between the semiconductor industry and the car industry and ensure a steady stream of innovation as well as assured spare parts supply.

1 Vehicle Architecture

1.1 Networking

Figure 1 shows a typical network of a high-end passenger car. The powertrain bus is realized with a high-speed CAN bus with 500 kbit/s datarate, the multimedia communication is assumed through the MOST-bus and the Body-network is built by the fault-tolerant low-speed CAN running at 125 kbit/s. In this architecture the LIN-bus acts as a sub-bus below the module level depicted.

In this architecture sensors are part of the functional subsystems like engine control, seat-control or navigation, and many more. In many control systems today, data is captured for the specific purpose of the subsystem. However it is much more practical, to make sensor data widely available. A good example is the speed information extracted from the four wheelspeed sensors of the ABS-system. In today's cars this speed information is distributed through the in-vehicle network to the engine management, the dashboard, the radio, and many other systems in the car may pick-up this data. In the synergetic case, the ABS system dispatches the consensed speed information on the network for the benefit of all bus participants.

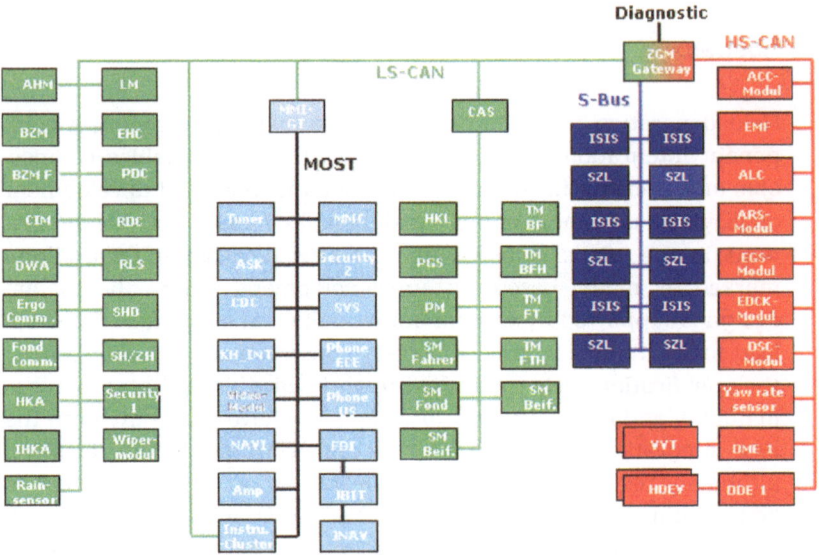

Fig. 1.   Example for a vehicle reference architecture.

The architecture developments in the car have engaged into major network reworks:

- ▶ In-vehicle networks are moving towards time-triggered architectures based on LIN™ and FlexRay™ for ease of system integration in a real-time signalling environment.
- ▶ Fault tolerant systems are emerging, based on safe networking, leading to higher system availability. System requirements for x-by-wire applications are being defined these days, early adopters will emerge on the market in 2006.
- ▶ Decomposition of functional system blocks into smaller entities is ongoing. The governing factor is the mechatronics concept of distribution and standard interfaces, which leads to standardization at the module level. This concept is novel to the car industry, which is still marked by a high number of customized electronics today.
- ▶ Concentration of computing performance in a few powerful nodes in the car is the architectural dream of the car industry today. In combination with a strong network, functional computations can then be moved around the network as per resource availability and ease of sys-

tem partitioning. The AutOsar initiative of the car-OEMs launched in September 2003, has the goal to define the software interfaces to make this software exchange possible.

1.2 Mechatronics

Flexibility in software module exchange goes hand in hand with hardware compliance to specific interface standards. The mechatronics concept defines the interfaces of combined mechanical actuators or sensors together with its controlling electronics. This entity is then encapsulated and ready for use in a larger vehicle context of networked modules. In this sense the software exchange concept in definition with AutOsar is linked intimately to the mechatronics hardware approach based on standard interfaces.

In this environment, components can then be sourced "Off-the-Shelf", discoupling its technological development from the semiconductor growth pattern. This process is of utmost importance for the fast moving electronics industry, as well as the slower automotive industry. It solves the lifetime discrepancy problem between electronics and automobiles. In a well designed system, spare parts can be sourced out of current production rather than from shelved components. Under this regime, products don't have to be produced in obsolete semiconductor lines or from components stored for many years under nitrogen any more. The mechatronics method is similar to the replacement of a light-bulb: old light bulbs are replaced with ones from current production, using newest technology rather than with stored bulbs, produced at the time when the car was build.

The rapid electronic progress makes semiconductor components obsolete much earlier than the long lasting car requires, which typically are between 15 and 25 years. In addition, the speed of innovation increases. In the mechatronics environment, new semiconductor technology can be applied faster to the application: as the mechatronics components have well defined boundaries, new components can be incorporated in the vehicle much faster. The resulting steady introduction of new semiconductor technologies helps speeding up innovation in cars additionally. This industrial mechanism copies the momentum in place with the computing and networking industries today.

However the status of mechatronic interfaces is not satisfactory yet. Although major interface standards exist on the signal level such as LIN™, CAN or FlexRay™, other mandatory interface standards are still desperately needed. Connectors exist in a large variety at different car makers, preventing significant exchange of components. On the same token mechanical fixtures are far from being standardized. In total the mechatronic concept is still far away from reality, due to the fact that these simple, but important interface standards are simply not existing industry-wide. Therefore the product example in figure 2, which is in production since a few years, will remain a single solution, if the industry cannot act quickly to set the mechanical requirements for mechatronics interfaces.

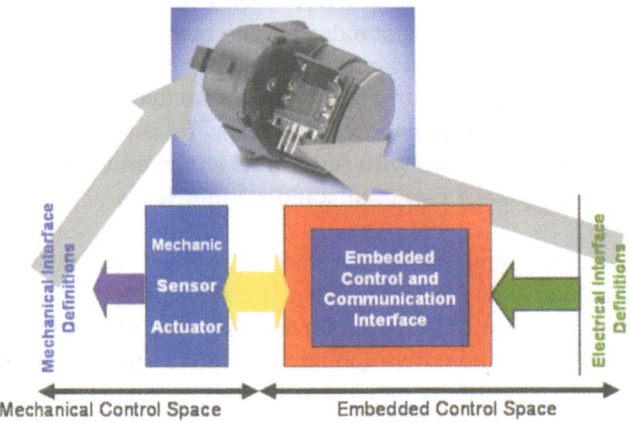

Fig. 2.   The mechatronics concept of distributed control.

1.3   Software Integration

When interfaces are fixed, the basic functions of the mechatronics components are too. In this context, innovation in functions is mainly driven through the software integration process. For this reason the interest in software development has drastically increased in the car industry over the last years. On the other side, in the mechatronics scenario, cost reduction is driven by producing these components cheaper when using new innovative production technologies or higher integrated electronic technologies.

Therefore the highest amount of innovation is directly linked to the system integration process. Here the main charter is to compose new system functions out of existing mechatronics modules. Mastering network technology is at the source, but more important are all aspects of software portability, which are consequently addressed by the newly formed AutOsar-consortium, where the middleware linkage layer standards are in the process of being defined.

In this context the middleware can be seen as an additional software bus which allows sharing of resources through standard exchange mechanisms (figure 3). At the same time hardware abstraction occurs in support of the mechatronics concept, thus establishing the basis for software portability over different computing and I/O platforms. In practice it will not matter any more if a PowerPC or a HC12, an ARM processor or any other controller core executes the code, as long as the timing and speed requirements are met. On the same token, bus access will be performed through standard routines, which will become part of the operating system, very similar to what the VOLCANO system from VolcanoAutomotive is already offering today.

A car system, operating under a strong network regime, equipped with a middleware system described above, will be marked by a drastic decrease of the number of ECUs employed. This meets the requirements of the car makers, who must reduce the number of nodes open for after-sales service with software maintenance substantially, in order to meet the quality targets of these new complex software cars.

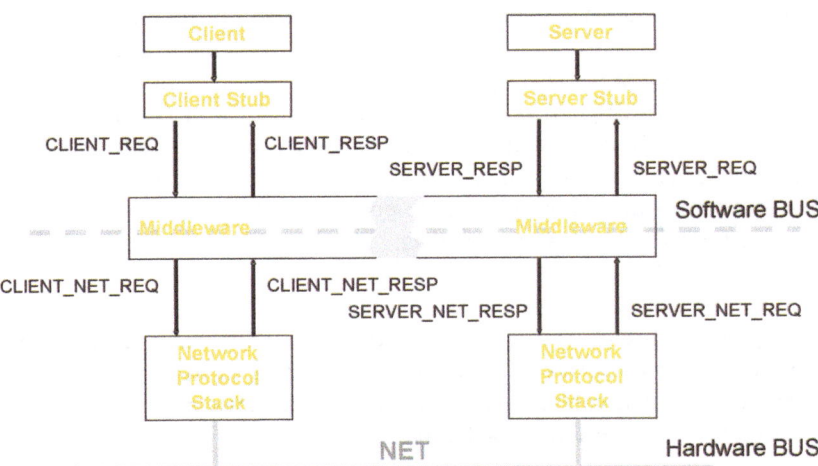

Fig. 3.   Middleware concept for mechatronics.

2   Sensor Technologies

The progress of MEMS surface micromachined semiconductor technology has led to rapid growth of the entire sensor market. The benefit of the semiconductor technology in this sector is clear: miniaturization means less material cost. The progress of semiconductor technology is matched into this domain, linking the shrink roadmaps of the semiconductor industry into sensing, too. The compatible semiconductor manufacturing technologies lend themselves to further monolithic integration of MEMS devices with its associated data treatment circuits, including the prospects of further cost and size reductions, reliability increases and functional enhancements. For this reason we can expect, that silicon based surface micromachined MEMS-technologies will quickly replace all other known sensor technologies in the future such as bulk-silicon machining or otherwise mechanically treated materials.

2.1 Motion Sensing

Motion sensing has been pioneering MEMS technologies. Its functional principle is a suspended polysilicon layer which acts as a capacitor. Its movement translates into a capacity change, which is being measured to yield the desired output of acceleration.

Motorola has developed a dual-die solution mounted on the leadframe before transfer mold into a standard plastic package. The sensor is shielded by a glass-frit from exposure with the plastic during mold. The electronic circuit is placed adjacent to the sensor die on the lead frame and performs the data treatment and calibration of the capacitance measurement as a function of the acceleration. It is capable to generate a DC-voltage for electrostatic deviation of the measuring plate for diagnostics purposes. Final test can thus be performed without mechanical movements.

The sensor is used in high quantities for airbag applications. Today, the output of the signal treatment circuit is ratiometric for further capture from an A/D channel of a microcontroller. In view of the anticipated architecture changes for stronger digital in-vehicle networking depicted above, digital interfaces are considered for future product derivatives.

2.2 Pressure Sensing

The underlying technology for pressure is similar to motion sensing. Through a hole in the package, the polysilicon layer is exposed to air-pressure (or other media), thus modulating the distance between another fixed layer according to changes in air-pressure. Again the transversal deviation of the material is first measured capacitively and then calibrated and amplified through an additional circuit, which finally generates the signal output for further digital processing by a microcontroller.

2.3 Imaging

Image sensing is moving rapidly towards static CMOS sensing, compatible with standard CMOS-technology, and away from dedicated CCD-technology. In combination with the available embedded computing performance in the 1000 MIPS-range, real-time object recognition becomes feasible for automotive traffic situations. Again the link to standard semiconductor technology makes this technology increasingly cost effective for general use.

2.4 RF technologies including Ultra-Wide Band

Radio frequency technology is receiving a new impulse through the advent of high bandwidth Si-Ge technologies. The 77-GHz band is in reach of monolithic technologies and permits to cost-effectively set up RADAR systems for advanced cruise control systems. RADAR is able to reliably determine distance and relative speed of other objects on the road and can be used in conjunction with imaging to strengthen the signal capture process for the car in traffic. In addition the traditional function of RF as a transmission medium for communication channels will rapidly increase once local car-to-car communication will receive attention for increasing road safety. Derived from the spread spectrum technology, broad GHz-bands are being assigned in some regions of the world for the two major applications: data transmission and Radar. The advantage of Ultra Wide Band (UWB) is that it can be configured for ultra-low power when only small transmission bandwidth is required.

2.6 MEMS based Initiators

There are numerous other applications for micromachining technologies. Today's airbag initiators are using HfHx-deposition aligned with semiconductor process technology. The integration of the initiator, with its driving electronics follows the mechatronic systems trend.

2.6 Data Fusion

Kalman filtering is an old concept to extract system state variables through measurement of peripheral system values and its transformation with estimated system properties. The same can be applied in modern cars. The many different measured variables can be transformed into a rugged view of the complete system, where the internal state variables are the result of a consensus process. All which is required is the knowledge of the system signal transfer properties and enough computing performance. As such, the capability is there today, and it should be emphasised that the massive increase of computing resources in the car will ease data fusion as a means of generating further internal signal data points.

3 Ambient Intelligence

With the availability of pervasive sensor data in the car, a new phase is about to start in automotive electronics. In combination with networking and massive computing resources, new systems are in definition which make traffic inherently safer, make travel times more predictable and driving more economic. Public authorities have recognized that the high number of fatalities on European roads must be reduced and have discovered that electronics still bears a largely underexploited potential. For this reason e-Safety initiatives concentrate on the capture of in-vehicle data as well as surrounding environmental data. Connectivity between cars and share of captured data uses the network effect of multiple data sources, leading to very economic safety systems in the car. The problem to be solved is not of technological nature, but key to success and widespread deployment is interface standardization. I.e. in which format data is to be shared between cars. In the jungle of multiple exchange mechanisms already introduced today, this is a very difficult problem to solve, especially as each proposed method has serious technical implications for the companies driving their preferred solutions. The reward however is big for the society: a phenomenal increase in transportation safety for the consumers and a new surge in semiconductor demand for the participating electronics suppliers.

4 Integration Process

The model based development concept is taking shape in the market place. It acknowledges the importance of the product design process for further improvements in product and system quality and places the focus on shifting development efforts away from the qualifying back-end into the "correct-by-design" front-end. However this requires a massive shift in development methodology through concurrent modelling and simulation efforts. The availability of models before the validation phase, will incur a major shift in the industry and can only be achieved as a concerted activity between car-OEMs, as well as its Tier-1 and Tier-2 suppliers.

4.1 Increasing product quality

The car electronic reports high component quality between 1 ppm and 10 ppm. Typically its component quality is higher than for other commercial industries, namely consumer and computer. However due to the lasting nature of cars and because failure of electronics is strongly impacting the end customer, quality needs to be improved beyond this level. Today's quality system relies on certification procedures at the end of the development cycle through summer- and winter-testing in the car. After this, the production process must be "frozen" to maintain this certified production process. The result is a rugged, but inflexible production process, with significant cost associated.

To go beyond this level, significant changes need to happen:

▶ The manufacturing volumes need to be increases by another order of magnitude. Serious learning of quality defects can only happen with volumes beyond 10 Mio pcs per year. For a 1 ppm failure rate, this means for this example that only 10 failing pieces are generated per year, which is not enough for deep rooted analysis.

▶ Quality is a component property, which must be designed-in rather than tested-in. This requires a paradigm shift in development, where certified development processes are used in the future (c.f. ISO-61508)

▶ On-Line component diagnostics need to be improved further. As the component is in use in the application on-line testing will assure that no degradation of system properties will occur. This property must be designed into the component. An option is to switch the component to service mode at specific predefined idle times.

▶ The software development and integration process must become as rugged as hardware design, where 1 Mio lines of VHDL-code can be handled with no failures. The good results in hardware design are due to the rugged encapsulation and reuse standards, which must be applied in a similar fashion to software.

▶ Encapsulation of generic (soft- and hardware) modules which can be qualified independently from the car will become a must do!

These are revolutionary changes in the way we write software and the implications are serious. However, it is well recognized, that the soft and hardware integration process must be improved significantly, or otherwise the product quality cannot be mastered in the next generation of complex automotive electronics.

4.2 Encapsulation

Mechatronics means encapsulation: A specific function is captured in a "box" and described through its interfaces. This is true for a mechatronics stepper motor with LIN-Bus capability as well as for a bus-capable temperature sensor. The same concept applies to embedded software in the same way. Today, UML is gaining acceptance in the automotive industry in order to increase reuse of existing code and master the complexity better through clean partitioning and solid description of the interfaces. Together with clean program management tools, the development process becomes more reliable and straight-forward than before. It aligns with IEC-61508 and lends itself easier for certification of safety critical components.

As a result the development resources can be shifted away from the "back-end", where product system integration testing, qualification and reliability engineering is done, towards the well-ordered development front-end.

This new design concept supports virtual prototyping and simulation, when carefully tailored interface layers separate the system into simulatable partitions. Practically, this means, that the industry will deemphasize investments into quality systems like QS9000 (now integrating into TS16949) and shift this efforts into design quality systems such as CMMI. In conjunction with a rugged and well followed development process, quality of the results can be guaranteed as well as project timeliness and budget compliance.

4.3 Tools

Encapsulation means increased model based developments, where for every encapsulated module we have the following properties:
- ▶ A full executable specification exist for every encapsulation, this can be a 'C'-model or Matlab model or any other executable description.
- ▶ A model of the environment is existing for the encapsulated block, preferably in the same description language.
- ▶ Simulation capability exists to exercise the model, which is virtual for hardware and "real" for software, to ensure operation.

The test-suite is defined and guarantees operation and qualification. The goal is to be able to qualify encapsulated component within its defined boundaries. With this, a car development becomes a layered approach with the primary objective of interface design, definition of the control hierarchies, and specification of the functionalities of the encapsulated module.

A large variety of development tools is on the market. An attempt is made here to order those into major categories. In brackets are some examples of existing tools and vendors:

- ▶ Requirements Management (Doors)
- ▶ Configuration Management (Clearcase)
- ▶ Defect Tracking (Clearquest)
- ▶ Process Flow Control (Rational Unified Process)
- ▶ Software Architecture (UML)
- ▶ Component Design: Case tools and Synthesis (Matlab Simulink, Stateflow, Veristate)
- ▶ Network Design (Volcano-Automotive, Vector, IHR)
- ▶ Simulation & Modelling Framework (System-C)
- ▶ Unit and Integration Testing (LDRA, PXI)
- ▶ Debug and Calibration (Metrowerks, Windriver, Lauterbach, PowerNexus)
- ▶ In essence a development team has to make a choice for a specific tool suite and work on the tool-interfaces, as a completely seamless environment is not yet in existence today.

5 Summary and Outlook

We have to recognise that the semiconductor microsystem technologies have led to a big increase of sensor data captured in the vehicle, making cars safer, more reliable and of higher use-value. This trend will continue as long as the sensor technology can profit from the tremendous semiconductor productivity increase, according to Gordon-Moore, which we continue to observe.

As an additional element of growth, the sensor data distribution will reach beyond the network of the vehicle. Progress in wireless technologies will ease capturing of data from further sources not accessible until now. Especially the treatment of environmental data around the car will revolutionize car safety systems. Data will be sourced from other cars, from sensors on the road infrastructure, or from a central server. This can be information about the actual road conditions, weather influences or particular traffic situations.

With cheap communication technology available, mobile ad-hoc networks will form on the road and between traffic participants to transmit pervasive sensor data through the system. New applications based on this concept of "Ambient Intelligence" will open the next chapter of exponential growth for microsystems and with this will generate a massive pull for new embedded semiconductors in the car.

To enable the integration process of Sensors in the network, encapsulation will be of utmost importance to comply with the mechatronics concept of automotive electronics architecture. General availability of sensor models will change currently existing development paradigms towards a model based automotive development system.

References

[1] Gerd Teepe: "Sensors in the next Generation Automotive Networks" Editor: Sven Krüger, Wolfgang Gessner, AMAA-Berlin, March 2002, Publication: Advanced Microsystems for Automotive Applications Yearbook 2002, pp276, ISBN 3-540-43232-9, Springer Verlag Berlin, Heidelberg, New York, 2002

[2] Gerd Teepe: "Elektronik im Lebenszyklus eines Kraftfahrzeugs" VDA-Technischer Kongress 20-21 März 2002, Stuttgart Liederhalle, Tagungsband:VDA, Verband der Automobilindustrie, Westendstraße 61, D60325 Frankfurt, Germany

[3] Christopher Temple, Gerd Teepe: "Composable Architectures" Elektronik Automotive, Magazin für Entwicklungen in der Kfz-Elektronik und Telematik, April 2002, WEKA Fachzeitschriften Verlag GmbH, Gruber Strasse 46a, 85586 Poing, Germany ISSN 0013-5658

[4] Gerd Teepe, Thomas Böhm: „Ambient Intelligence - Ein Konzept für den unfallfreien Verkehr der Zukunft" Automotive Electronics Sonderdruck II/2002

[5] Gerd Teepe, Don Remboski, Richard Baker: „Towards Information Centric Automotive System Architectures" Paper Number SAE 2002-21-0057, Proceedings of the 2002 International Congress on Transportation Electronics, October 21-23, 2002 Cobo-Center, Detroit Michigan;ISBN 0-7680-1113-2, ISSN 98-86269, SAE/P-381 SAE Customer Service email: CustomerService@sae.org

[6] Gerd Teepe: „Schnelle Innovationszyklen in der Halbleiterindustrie im Kontrast zu langlebigen Produktstrategien der Anwenderindustrien" 6. Handelsblatt-Jahrestagung HALBLEITER-INDUSTRIE 2002 18-19 September 2002, Kempinski Hotel Atlantic, Hamburg

[7] Gerd Teepe: „Open Assembly Architectures for flexible Plug&Play Car Solutions" MI-Congress 2002, „Automobile Electronics, From Systems to Vehicle Networks, from Semiconductors to Software" Liederhalle Stuttgart, Germany, November 26-27, 2002 Verlag Moderne Industrie, 56895 Landsberg am Lech

[8] Can in Automation (CIA) webpage on: www.can-cia.org

[9] LIN-Consortium; "LIN-Specification Version 1.3" webpage on: www.lin-subbus.org

[10] FlexRay Consortium webpage on: www.flexray-group.org

[11] AutOsar Consortium webpage on: www.autosar.org

[12] Gerd Teepe; Thomas Görnig: „Automotive Sensor Integration", Proceedings of the AMAA 2003, Advanced Microsystems for Automotive Applications 2003, edited

by Jürgen Valldorf and Wolfgang Gessner, Springer Verlag ISBN 3-540-00597-8, pp. 509-518

[13] Gerd Teepe; Bernd Rucha: „Technological Requirements for Networked Traffic – A future e-safety concept", 9th EAEC International Congress „European Automotive Industry Driving Global Changes" 16-18 June 2003, Paris, France, Palais des Congrès, Proceedings by: Usine Nouvelle, 12-14 rue Mederic, 75815 Paris Cedex, France, www.usinenouvelle.fr

[14] Gerd Teepe: „FlexRay-Kommunikation und Systemintegration in der nächsten Generation Automobilelektronik" Euroforum Konferenz „Datenkommunikation im Automobil" 3-4 Juni 2003, Heidelberg, Proceedings EuroforumDeutschland GmbH, Prinzenallee 3, 40512 Düsseldorf

[15] Gerd Teepe; Bernd Rucha: „Technological Requirements for Networked Traffic – A future e-safety concept", 9th EAEC International Congress „European Automotive Industry Driving Global Changes" 16-18 June 2003, Paris, France, Palais des Congrès, Proceedings by: Usine Nouvelle, 12-14 rue Mederic, 75815 Paris Cedex, France, www.usinenouvelle.fr

[16] VolcanoAutomotive, Company to provide Network Integration software and operating Systems. www.volcanoautomotive.com

Gerd Teepe
Motorola Semiconductor Products Sector
Schatzbogen 7
81829 München
Germany
gerd.teepe@motorola.com

Keywords: Mechatronic, Automotive electronics Architecture, plug & play, lifetime supply, composable architectures, mechatronic standards, Autosar, LIN, CAN, FlexRay, Osek, Model based system development, integration quality

Networking In-Vehicle Entertainment Devices with HAVi

A. Leonhardi, T. Gschwandtner, M. Simons, DaimlerChrysler AG
V. Vollmer, G. de Boer, Robert Bosch GmbH

Abstract

Modern in-vehicle telematics systems typically integrate a broad range of entertainment features, like AM/FM radio, TV, and CD/MP3/DVD players, with advanced navigation and communication features and present them to the user through a consistent automotive-adequate interface. In the future, passengers will be able to access and control these features independently through their own consoles. For the required high-speed automotive-grade internetworking, the industry has established the MOST (Media Oriented Systems Transport) standard. This standard specifies the physical layer up to the network management layer together with a set of application oriented device interfaces. For distributed applications, however, the standard lacks an appropriate middleware layer. HAVi (Home Audio/Video Interoperability) is a carefully designed middleware standard that allows IEEE 1394 (Firewire) based devices to interoperate. In this paper, we show that HAVi can be adapted to the MOST bus. In particular, we describe in detail the essential changes to the HAVi standard that are required to use HAVi as a middleware layer on top of MOST.

1   Introduction

The automotive industry is currently in a transition phase from being hardware focused to being software driven. A modern premium vehicle is a complex distributed software intensive system on wheels involving several million lines of code and interconnecting as many as 77 control units by means of a hierarchy of domain-specific heterogeneous sub networks (see also [3]). The challenges faced when engineering these advanced vehicles rival those of modern airplanes. One sound approach to meet these challenges is to develop middleware standards that provide a stable programming interface for applications. One example is the OSEK/VDX family of standards for an open-ended architecture for distributed control units in vehicles [9].

In-vehicle telematics systems are themselves sophisticated distributed soft-ware intensive systems (cf. figure 1). They typically integrate a broad range of entertainment devices, like AM/FM radio, television, and CD/MP3/DVD play-ers, with advanced navigation and communication features and present them to the user through a consistent automotive-adequate interface. In the future, passengers will be able to access and control these features independently through their own control units. For the required high-speed automotive-grade internetworking, the industry has established the MOST (Media Oriented Systems Transport) standard [6]. The MOST bus provides the means for trans-mitting control commands, asynchronous messages and synchronous multi-media streams with an overall bandwidth of 25 Mbit/s. It interconnects con-trolled (e.g. a CD changer) and controlling multimedia devices (the head unit and rear seat control units). The standard specifies the physical layer up to the network management layer together with a set of application oriented device interfaces. For distributed applications, however, the standard lacks an appro-priate middleware layer including higher-level management services (such as resource management, event management, stream management, or a general bus-independent messaging). These services typically would have to be devel-oped in a proprietary manner.

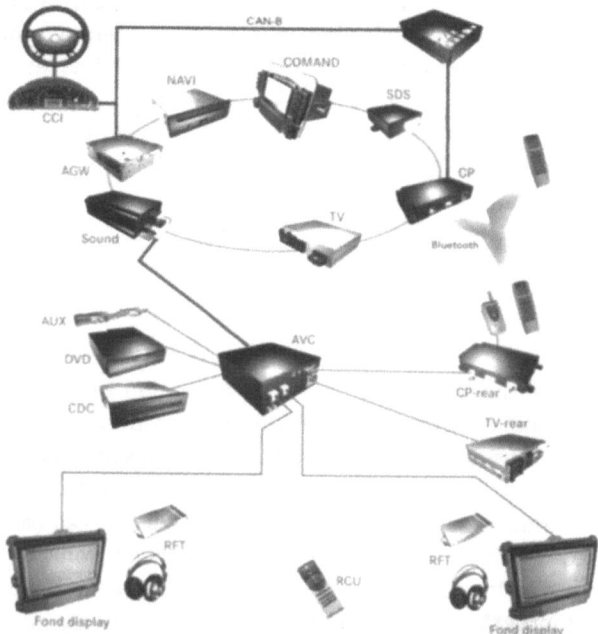

Fig. 1.   MOST based telematics ring in the Maybach.

HAVi (Home Audio/Video Interoperability) is an existing carefully designed middleware standard that allows IEEE 1394 (Firewire) based devices to inter-operate [1, 10]. In this paper, we show that HAVi can be adapted to the MOST bus. The advantages are twofold: first, we cleanly separate the application logic from communication / transport / media access aspects, and, second, we reuse the system management services of the HAVi middleware. We summarize the essential changes to the HAVi standard that are required to use HAVi as a middleware layer on top of MOST. These changes are documented in detail in a draft specification [11]. The guiding principle has been to follow HAVi's design principles and to keep changes as minimal as possible. Therefore, these changes could be easily integrated into a future version of the HAVi standard.

One of the original design goals of HAVi has been to be transport layer inde-pendent. The present HAVi specification, however, does not meet this goal completely. IEEE 1394 specific issues show up in layers that should in princi-ple be shielded from them such as addressing, stream management, and device discovery. One challenge of adapting HAVi to MOST therefore was to identify these areas and to clearly factor out transport layer specific aspects. This is a further – more general – contribution of our work and could be used to develop a future transport layer independent version of the HAVi standard.

The remainder of this paper is structured as follows: The next section shortly describes the HAVi standard and how it could be applied in the automotive environment. Section 3 describes the MOST bus and the changes necessary to extend the HAVi standard to support the MOST bus. In section 4 related work is discussed while section 5 concludes the paper.

2 Utilizing the HAVi Multimedia Framework in the Automotive Environment

The HAVi architecture (defined in [1]) has been developed to allow for an inter-operability and extensibility of devices in the home entertainment environ-ment, which is in many aspects similar to the entertainment networks in vehi-cles. In this section we first describe the HAVi framework and give examples of possible uses in a vehicle entertainment system. We then discuss changes necessary to the HAVi standard because of the special requirements of the automotive environment.

2.1 The HAVi Standard

The HAVi standard specifies a set of distributed management services and their interfaces as required in the home entertainment environment, for example for the management of video and audio streams. Furthermore, a controlled device in the HAVi environment may provide its own device driver (called a Device Control Module, DCM), which can be executed on a controlling device, and its own user interface (in a so-called Havlet). This concept makes a HAVi network extensible, because it allows to add previously not supported controlled devices.

An important aspect of the HAVi architecture is the distinction between different types of devices, which allows to have complex controlling devices (e.g. set top boxes) as well as more simple and therefore cheaper controlled devices (e.g. a CD player). Additionally, this concept allows to integrate legacy devices that are not aware of HAVi. Both is important for the automotive environment, as the former allows to include more of the numerous often simple Electronic Control Units (ECUs) in the HAVi network and the latter because it allows a reuse of existing devices, reducing the transition costs.

The HAVi architecture defines the following device categories:

- ▶ Full Audio/Video device (FAV): FAVs are controlling devices that run all HAVi managers (see below) and are able to execute the DCMs for controlled HAVi devices (BAVs and LAVs). To run an uploadable DCM provided by a BAV, an FAV has to contain a runtime environment for Java byte code. Applications using the HAVi framework will be executed on an FAV or IAV. The head unit of a vehicle telematics system would be an FAV.
- ▶ Intermediate Audio/Video device (IAV): An IAV is similar to an FAV but has a reduced set of managers and no runtime environment for Java byte code. This allows for cheaper controlling devices with less resources. In the automotive domain, a rear-seat unit could be realized as an IAV.
- ▶ Base Audio/Video device (BAV): BAVs are controlled devices that provide HAVi-specific information and a DCM code unit as uploadable Java byte code in its ROM. An FAV can download this code unit and run the DCM for a BAV.
- ▶ Legacy Audio/Video device (LAV): Existing controlled devices that are not aware of HAVi can be included in a HAVi network as LAVs. An FAV or IAV can provide an embedded code unit and run a DCM for such a device (e.g. a device supporting the IEEE 1394-based Audio/Video Control protocol, AV/C). Existing controlled MOST devices, such as a CD changer or a radio tuner, will have to be treated as LAVs.

The management components defined by the HAVi architecture are shown in figure 2 for a HAVi FAV device. The Communication Media Manager for the IEEE 1394 bus (1394-CMM) provides an abstraction for sending and receiving data via the IEEE 1394 bus. The Messaging System provides a higher-level communication mechanism where applications and managers on different FAV/IAV devices can communicate either by exchanging messages or by using a Remote Procedure Call mechanism called HAVi Remote Method Invocation (RMI).

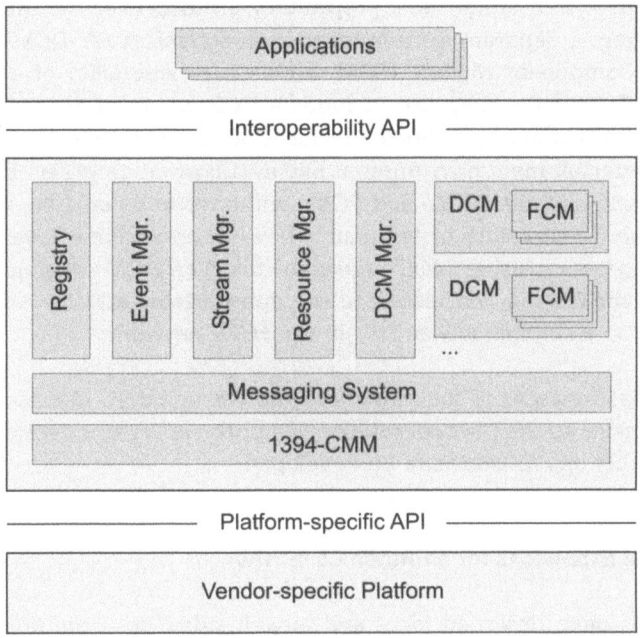

Fig. 2.   Components of the HAVi architecture.

With the Registry all applications and services (called Software Elements) have to register with describing attributes. Usually the telematics system of a vehicle comes in many different variants, for example with or without integrated mobile phone. An application will therefore use the Registry to determine what devices and services are available in the system. The Event Manager is responsible for propagating system events, indicating for example a network reset or new and gone devices, in the HAVi network. Connections for transmitting multimedia data between devices can be set up and controlled through the Stream Manager. In a vehicle telematics system multiple streams, for example for radio, telephone as well as warning signals, will have to be coordinated. The Resource Manager is responsible for managing conflicting access to a resource of the HAVi network. Using the Resource Manager an application can reserve a certain resource or start a negotiation process, if a resource it wants to use is already occupied. The Resource Manager would be used in a telematics system to manage the conflicting access for example to a CD changer from front- and rear-seat control units.

Each device in a HAVi network is represented by a Device Control Module (DCM). The DCM for an FAV or IAV runs on the device itself, the DCM for a BAV or LAV device on a suitable FAV or IAV. A DCM for an LAV device can communicate with it either via a proprietary protocol over the IEEE 1394 bus or even using a separate communication mechanism. A DCM provides a Functional Component Module (FCM) for each functionality of a device. For example, a TV system that also includes a VCR would provide two separate FCMs. The HAVi standard defines interfaces of a number of typical devices for the home entertainment environment like a VCR or an amplifier. For the automotive domain suitable DCMs and FCMs will have to be created, which allow to access the functionality of telematics devices on a higher level of abstraction as compared to using MOST mechanisms. The DCM Manager, finally, is responsible for finding and installing the appropriate DCM for each BAV and LAV device on a suitable FAV or IAV in the HAVi network.

Together the interfaces of the managers and the standard interfaces of DCMs and FCMs make up the Interoperability API of the HAVi architecture, based on which portable applications can be developed.

2.2 General Extensions for an Automotive HAVi

As HAVi has been designed for – and is well suited to – the home environment, some modifications and extensions are becoming necessary for the use in the automotive environment. While the basic requirements at home are easy installation and use, in a vehicle they are system stability and operation in rough environments. This leads to the following extensions of the HAVi specification, which are discussed in more detail in [8]. An important guideline for the design of these modifications was the downward compatibility with the existing HAVi specification and devices:

▶ **Power Consumption Management** – differently from the home environment, the power consumption of the devices is a crucial issue in automotive networks. We therefore propose to add a network-wide power management to the HAVi architecture. It consists of a Power Management Server (PMS), that operates as a HAVi manager, and a number of Power Management Clients (PMCs) representing the HAVi-based devices in the network. Every decision on changing from normal activity to a lower activity level (as down to deep sleep mode) is taken by the server and can be based on a large variety of information sources and pre-configured scenarios.

▶ **Network-wide Error Handling** – to cope with the increased demands on system stability, an error handling will have to be added, which has far-reaching capabilities to detect errors and to react accordingly. Similar to the power management, the error management has one central element – the Error Management Server (EMS) and distributed Error Management Clients (EMCs) that support and inform the EMS to detect errors and to perform selected exception handling mechanisms locally.

▶ **Smart Device Access Management** – very important in distributed networking is the management of competing device or function accesses (i.e., two applications residing in the network try to access the same network resource). HAVi has sophisticated means to deal with these scenarios, but relies heavily on user decisions to resolve access conflicts. In a vehicle this is not acceptable because of driver distraction and will have to be replaced by a mechanism, which considers applications' priorities. Based on these priorities an automotive HAVi can ensure deterministic access for multiple applications to the same resource.

▶ **Smart Bandwidth Management** – having the access to a certain resource might not be sufficient, if the required data transmission rate is not available. As in HAVi multimedia streams are already controlled through the Stream Manager, this component can be modified to take into account the same priorities as for the device access management and to release or to suspend connections accordingly.

As mentioned before, HAVi provides standardized FCM interfaces for the most common home devices. Because vehicle telematics devices are very specialized and differ from home devices, FCMs for common vehicle devices need to be added. The following shows some examples for automotive FCMs, which will be based on the corresponding MOST function blocks to a large degree:

▶ **AV Disc Changer FCM** – extends the standard AV Disc FCM with the yet lacking functionality to change the active disc.

▶ **Navigation FCM** – provides access to the navigation unit, for example for entering a destination automatically.

▶ **Automotive Tuner FCM** – comprises support for traffic announcements and other automotive specific needs for tuner applications.

▶ **Mobile Phone FCM** – will allow the control of a mobile phone.

3 Adapting the HAVi Framework to the MOST Bus

In the previous chapters we have discussed the requirements of the automotive domain regarding a middleware for its multimedia applications. We have also shown how the HAVi framework fulfils these requirements. In this chapter we discuss the changes necessary for adapting the HAVi framework for use with the MOST bus. Because the MOST bus is currently coming into use in vehicles such as the Audi A8, the BMW 7er series and the Mercedes E-Class, adding support for MOST to the HAVi standard is crucial for its use in the automotive environment.

3.1 The MOST Bus

The MOST bus (as defined in [6]) provides the means for communication between controlling and controlled multimedia devices and is able to transport multimedia data in a synchronized way. It has been designed especially with the requirements of the automotive environment in mind. Because the MOST bus transmits data in a synchronous way and because of its low susceptibility to electro-magnetic interference, it can provide the required reliability.

The MOST bus is structured in a ring topology. In its current version, it supports up to 64 devices and provides an overall available bandwidth of 24.8 Mbit/s. It uses plastic optical fiber (POF) as its connection technology, which results in a reduction of the problems caused by electro-magnetic interference, but is more robust than glass optical fiber.

For transmitting data, the MOST ring supports the following types of data channels with different characteristics:

- ▶ **Control channel:** The control channel is used to send control messages to a device in order to initiate a certain functionality (e.g. starting to play a certain track of a CD) or to retrieve information about its current state (e.g. the track position). To this end, control commands and their parameters as well as the device's responses or periodic status reports are exchanged via control messages. The control channel provides a bandwidth of 705.6 kBit/s.
- ▶ **Synchronous channel:** For transmitting multimedia information between devices in a MOST ring, the MOST bus supports up to 60 physical channels, where a physical channel corresponds to one byte at a rate of 44,1 kHz. Physical channels can be grouped to form a logical channel; for example, it takes four physical channels to make up one stereo audio channel of CD-quality. The MOST bus can therefore support a maximum of 15 such stereo channels or 21.2 Mbit/s.

▶ **Asynchronous channel:** The asynchronous channel is intended for transmitting larger amounts of data in a more burst-like manner. Access to the asynchronous channel is provided through a token mechanism. Of the maximum number of 60 physical synchronous channels (see last point), the capacity of between 0 and 36 can be allocated to the asynchronous channel. This allocation can be parameterized by setting a so-called boundary descriptor according to the needs of the applications. If the full complement is allocated - leaving 24 physical synchronous channels - 12.7 Mbit/s are available on the asynchronous channel.

The distinction between the control and the asynchronous channel for data communication allows for simple controlled devices (such as a CD player) that do not support asynchronous communication, making hardware requirements for these devices less expensive.

Besides the low level communication mechanisms, the MOST bus also provides mechanisms for controlling the functionality of devices in a MOST network. A certain functionality is encapsulated in a function block, whose purpose can be compared to that of a HAVi FCM. The AudioDiskPlayer function block, for example, represents the CD player of the vehicle's sound system. Via the MOST bus an application can control the function blocks available in the MOST ring and receive information about their status. MOST specifies standard function blocks for accessing typical functions of an automotive telematics system such as a CD player, an amplifier or a mobile phone.

3.2 Main Differences between HAVi and MOST Mechanisms

Because the HAVi architecture is influenced by the underlying IEEE 1394 bus, there are the following differences to the mechanisms of the MOST bus, which will have to be addressed for an integration of support for MOST:

▶ As the network environment in a vehicle is usually static, the MOST bus only uses local addressing schemes (16 bit) based on a device type specific address (logical node address) or the position of the devices in the ring (node position address). There is no global unique addressing scheme similar to the global unique identifiers (GUIDs) used in an IEEE 1394 network. Additionally, addresses may change when the configuration of the ring is changed.

▶ Unlike the IEEE 1394 bus, where control messages are sent as regular messages of a special format, the MOST bus has a separate control channel for sending control messages with its own message format and mechanisms.

▶ The management of multimedia streams is more simple in the MOST bus as compared to the IEEE 1394 bus. The sources and sinks of a device are accessed only through the corresponding function blocks. HAVi on the other hand, makes a distinction between FCM plugs and device plugs and allows for internal connections. Also, in the MOST bus it is not possible to control the properties of a stream dynamically.

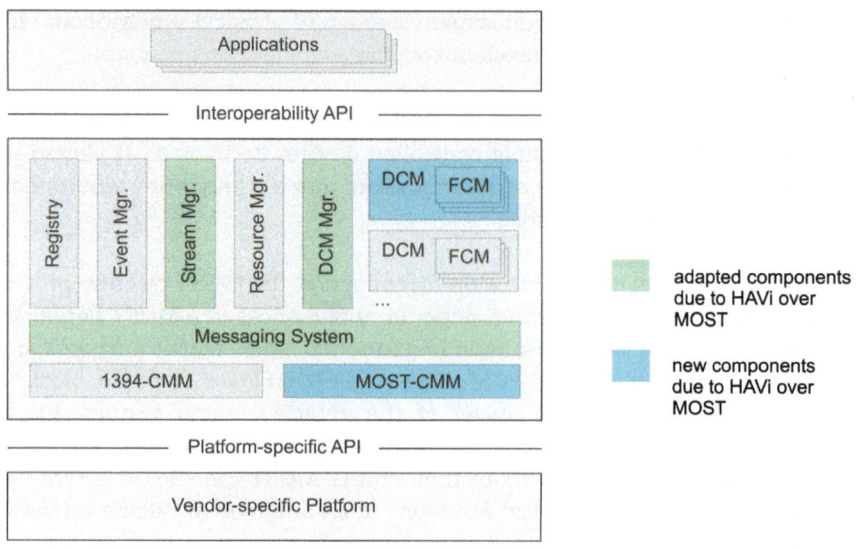

Fig. 3.   Changes to the components of the HAVi architecture (compare to figure 2).

To address these issues the following changes (cf. figure 3), have to be made to the components of the HAVi architecture. All other components not mentioned here are not influenced by the extension:

▶ The most noticeable change is a new MOST Communication Media Manager (MOST-CMM) that encapsulates all access to the MOST bus (i.e. all other components described below use the MOST-CMM to access the MOST's functionality). The MOST-CMM is described in detail in the following sub section.

▶ The interface of the Messaging System does not need to be changed. However, the Messaging System has to detect if a device in the MOST network is accessed and to call the MOST-CMM instead of the 1394-CMM. In a network which contains both a MOST and an IEEE 1394 network, the Messaging System additionally has to provide the functionality of a gateway.

▶ The functionality of the Stream Manager is based on the IEC 61883 protocol extension of the IEEE 1394 bus. The functionality of the Stream Manager therefore has to be adapted to take into account the different (more simple) stream management capabilities of the MOST bus.

▶ A Device Control Module (DCM) and its associated Functional Component Modules (FCMs) control the access to the functionality of a certain device. They therefore have to implement the device specific functionality of automotive MOST devices.

▶ Finally, the Device Control Module Manager has to be extended slightly to be able to detect MOST devices and to find and install suitable DCM code units.

3.3 The MOST Communication Media Manager

The MOST Communication Media Manager (MOST-CMM) is the only new component that has to be added to the HAVi architecture for a support of the MOST bus. We will therefore describe it here in more detail.

One important responsibility of the CMM is the mapping of the GUID addresses used in HAVi (which every IEEE 1394 device already provides) to the addresses used in the communication system. As the MOST bus provides no globally unique addressing scheme, we have to map MOST addresses to appropriate HAVi GUIDs. Based on this address the MOST-CMM in a gateway device has to be able to decide, to which attached communication bus a certain message has to be routed to.

We propose a simple solution (see figure 4), where the GUID begins as before with a leading OUID (Organizational Unit Identifier) describing the vendor who manufactured the device. To guarantee that an automotive GUID is different from the GUIDs of existing IEEE 1394 devices, a vendor that produces devices for the automotive environment will have to request a separate OUID for this purpose.

The remaining field of the GUID begins with a 1 byte identifier of the sub network, called the NetId. This identifier is different for each communication system attached to a gateway device. During initialization the NetId is 0 (which is also the default value, when no gateway functionality is required) and is then set by the gateway device appropriately. The next byte differentiates between MOST devices that are aware of HAVi (0x00) and devices that are not (0x01).

A MOST device that is HAVi-aware will provide a vendor-wide unique serial number, which constitutes the remainder of the GUID. The GUID for a MOST LAV device (which is not HAVi-aware) will contain its logical node address in the last two bytes. Finally, a further byte, the BusId, can be used to differentiate between different MOST buses that are directly connected, for example if there is a separate MOST bus for the front- and the rear-seat entertainment system.

automotive GUID

OUID (24 bit)	NetId (8 bit)	automotive vendor specific part (32 bit)

MOST LAV devices

OUID (24 bit)	NetId (8 bit)	reserved (8 bit) = 0	BusId (8 bit)	logical node id (16 bit)

automotive MOST HAVi devices

OUID (24 bit)	NetId (8 bit)	reserved (8 bit) = 1	serial number (24 bit)

Fig. 4.   Addressing scheme for MOST devices.

As mentioned before, we decided to encapsulate all access to the MOST bus in the MOST-CMM. This is somewhat opposed to the original HAVi specification, where higher level components need to have knowledge about IEEE 1394 specific protocols. The Stream Manager, for example, uses the IEC 63881 protocol [2] to manage multimedia connections. Therefore, besides an interface for sending and receiving data, the MOST-CMM also provides interfaces for retrieving information about devices and for managing multimedia streams:

- ▶ **Retrieving device information:** Besides the method CmmMost::GetGuidList, which enumerates the GUIDs of all devices in a MOST network, the MOST-CMM offers the method CmmMost::GetDeviceInfo to retrieve information about a MOST device, including vendor, product name and a list of the function blocks the device implements.
- ▶ **Asynchronous messaging:** We decided to use the asynchronous channel of the MOST bus to transmit HAVi messages (including HAVi RMI), as it is best suited for the messaging style of communication and unlike the control channel provides the necessary bandwidth. To this end, the MOST-CMM offers the method CmmMost::SendPacket and an appropriate indication mechanism similar to the original Cmm1394::Write.

► **Controlling MOST devices:** While asynchronous messages can be used for the communication between FAVs/IAVs, a different messaging mechanism is required to control existing MOST devices, which has to be done by sending control messages over the control channel. The MOST-CMM therefore offers similar methods as for asynchronous messages for sending and receiving control messages.

► **Managing synchronous connections:** Finally, the MOST-CMM supports various functions to manage synchronous connections and to query their current state. With CmmMost::Allocate a certain source plug of a MOST function block can be caused to allocate a synchronous channel and with CmmMost::Connect a sink plug to connect to this channel. The available bandwidth on the MOST bus can be queried with CmmMost:: AvailableChannels, information about the sink and source connectors of a certain MOST function block with CmmMost::GetSyncDataInfo.

3.4 Stream Management

Changes to the HAVi Stream Manager are necessary, because the stream management is more restricted in the MOST bus than with the IEEE 1394 bus. The main conceptual differences are as follows:

► While HAVi and IEEE 1394 make a distinction between device and FCM plugs, the MOST bus only supports connectors for function blocks, similar to FCM plugs. To remain compatible, a DCM for a MOST device will therefore support a virtual device plug for every FCM plug of its associated FCMs.

► In IEEE 1394 it is possible to connect FCM plugs on the same device without establishing a connection over the communication system. In the MOST bus, however, there is no general mechanism to establish such internal connections.

► In the MOST bus, the characteristics of a stream are defined by its source connector. It is not possible to set properties of the stream, like bandwidth, data type etc., dynamically. This restriction also has to be reflected by the Stream Manager.

3.5 Device Control

During start-up of a HAVi network, the DCM Manager scans the network for controlled BAV or LAV devices. For IEEE 1394 devices, all except LAVs can be identified through the information in their SDD ROM. The next step is to find an appropriate DCM for each of them. For a BAV this is done by uploading the DCM code unit either from the device or from a location it specifies. For an

LAV the DCM code unit has to be provided by the controlling FAV/IAV itself. If more than one FAV is able to run the DCM for a certain device, they initiate a negotiation process to determine the most suitable one.

If an FAV/IAV also supports MOST devices, it has to perform the same procedure for the devices in an attached MOST network. For a HAVi-aware MOST device, the function of the SDD ROM is supplied by a special MOST function block called HAVi (with the tentative function block identifier 0xD0). The Device Manager will then check if any of its embedded DCM code units matches a detected device and install this code unit as for an IEEE 1394 device (if necessary after an appropriate negotiation process).

In an object-oriented HAVi stack (e.g. the FireBus stack from VividLogic [7]) a concrete DCM or FCM is often derived from a generic parent class, called for example GenericDcm or GenericFcm, which provides general functionality, like registering the software elements with the Registry. For the integration of MOST devices in HAVi, we propose to encapsulate the generic functionality of the MOST bus in a second abstract layer consisting of the classes MostDcm and MostFcm. These are intended to make the development of MOST DCMs and FCMs considerably easier by already including generic functionality, which any MOST device has to implement according to the MOST standard (e.g. returning the product or company name). A specific DCM or FCM derived from these generic classes only has the responsibility to implement the special functionality of the device, for example that of a TV tuner. It can keep the generic functionality provided by the more abstract classes or extend it to be more specific, where necessary.

4   Related Work

Today, the different manufacturers of vehicle telematics components use proprietary software architectures (such as Siemens VDO's Top Level Architecture, TLA) to implement the devices' functionality. Because the architectures of devices from different suppliers have no common understanding of distributed management functions, they can only interact on the level of MOST mechanisms. As mentioned in the introduction, this means that few higher level management functions (like resource management) can be reused between following generations of a telematics system. Instead, they have to be defined and implemented separately for each generation.

Another standard that we have considered for connecting vehicle telematics devices in a flexible and extensible manner is Microsoft's Universal Plug and Play (UPnP) [12]. As the communication in UPnP is based on TCP/IP, it is more independent from the underlying communication network than HAVi. In UPnP device interaction is handled in a peer-to-peer fashion. The services offered by a device (actions or events) are described in a general way by a document of XML format. A service announces itself in the network by means of a service advertisement sent via multicast on a well-known address with a limited time to live (UPnP proposes a time to live of 4 hops). With a similar mechanism, an application can query for offered services through a service request. For the control of a service, control commands and events are exchanged via the Simple Object Access Protocol (SOAP). However, the mechanisms of UPnP are more suited for a flexible environment as compared to the more static environment in a vehicle. As yet, UPnP provides no mechanisms for a distributed management of resources. Interfaces to specific services are just beginning to be defined.

A further standard which defines the interaction of controlled and controlling devices on a higher layer of abstraction is Jini [13], which is based on Sun's Java mechanisms. Here devices announce themselves to a special Jini Lookup Service in their vicinity, again through a multicast with a restricted time to live. Applications query a nearby Lookup Service, found through a similar multicast-based mechanism, for a service they want to use. Services have to provide a proxy as Java byte code, which handles all interaction with the service and which can be retrieved through the Lookup Service. An application uses this proxy, which can be compared to a DCM code unit in HAVi, to access the service. As with UPnP, Jini is more suited for a flexible environment and does not provide the distributed management services required for the automotive environment.

5   Conclusion

In this paper, we presented the case for employing a multimedia framework to achieve extensibility and reusability of software in the automotive telematics domain. We showed the domain-specific requirements for such a framework and concluded that the HAVi framework, which originally has been developed for home entertainment systems, matches these requirements to a large degree. Based on this, we discussed necessary extensions of the HAVi architecture. During this discussion, we concentrated on describing the support for the MOST bus, which is beginning to become widely used in modern upper-class vehicles. Further extensions have been sketched briefly and are the focus of related work [8].

By extending the HAVi framework with support for the MOST bus, we also showed the possibility of a HAVi that is independent of the underlying communication bus. The changes necessary for a support of the MOST bus indicate the parts of the HAVi architecture that have to be adapted to add support for an arbitrary communication mechanism. This would also allow for a HAVi, which is for example based on TCP/IP. We believe that such an independence from the communication system could also significantly improve the use of the HAVi framework in the home entertainment environment.

References

[1] HAVi Inc.: The HAVi Specification, Specification of the Home Audio/Video Interoperablity (HAVi) Architecture. Version 1.1, May 2001, URL: http://www.havi.org.

[2] IEC: IEC 61883 Standard for a Consumer-Use Digital Interface, Parts 1-5.

[3] Leen, G., and Heffernan, D.: Expanding Automotive Electronic Systems. IEEE Computer, Vol. 35, No. 1, pp. 88-93, January 2002.

[4] Microsoft Corp.: Windows CE for Automotive, Web-Site, URL: http://www.microsoft. com/automotive/, December 2002.

[5] Mitsubishi Digital Electronics America Inc.: NetCommand® Technology, Web-Site, URL: http://www.mitsubishi-tv.com, December 2002.

[6] MOST Cooperation: MOST Specification, Revision 2.2, November 2002.

[7] VividLogic Corp., VividLogic Home Page, Web-Page, URL: http://vividlogic.com, December 2002.

[8] Vollmer, V., Robert Bosch GmbH: Supplement to IDB-1394: System and Interoperability Layer, Proposal to 1394 Trade Association, Automotive Working Group, July 2002.

[9] OSEK/VDX: Open-ended architecture for distributed control units in vehicles, Web-Site, URL: www.osek.org.

[10] Lea, R., et al.: Networking Home Entertainment Devices with HAVi, IEEE Computer, September 2000, pp. 35-43.

[11] DaimlerChrysler AG, Robert Bosch GmbH: HAVi over MOST Specification, Extending the Home Audio/Video Interoperability (HAVi) Architecture to Operate over the Media Oriented Systems Transport (MOST) Bus System, Version 0.9.4, December 2002.

[12] UPnP Forum: Universal Plug and Play Forum Homepage, Web-Site, URL: www.upnp.org.

[13] Sun Microsystems: Jini Homepage, Web-Site, URL: www.jini.org.

Alexander Leonhardi, Thomas Gschwandtner, Martin Simons
DaimlerChrysler AG
Telematics Research, RIC/TA
HPC T729
70546 Stuttgart
Germany
{alexander.leonhardi, thomas.gschwandtner, martin.simons}@daimlerchrysler.com

Vasco Vollmer, Gerrit de Boer
Robert Bosch GmbH
FV/SLH
P.O. 77 77 77
31132 Hildesheim
Germany
{vasco.vollmer, gerrit.deboer}@de.bosch.com

Keywords: in-vehicle telematics systems, multimedia middleware, MOST

Digital Audio Broadcasting – Solutions for the Automotive Environment

S. Evans, Frontier Silicon

Abstract

Digital Audio Broadcasting (DAB), based on the Eureka-147 specification is rapidly gaining a foothold in the consumer market, with an increasing number of broadcasters in many countries promoting the new services that DAB offers, as well as the excellent audio reproduction made possible through digital transmission.

Additionally, DAB offers significant extra capability for in-car infotainment systems from the large data capability supported by the Eureka standard. In the automotive environment this bandwidth can be deployed for traffic service information, using standards such as RDS-TMC and TPEG.

Also, emerging applications such as off-board navigation are very hungry for bandwidth, and with the right server technology in the broadcasting chain DAB offers a very credible alternative to the expensive cellular network for delivering large volumes of data to individual vehicles.

As the leading supplier of DAB solutions to the electronics industry, Frontier Silicon has also developed automotive DAB products, chipsets and modules, that enable 1st tier suppliers to quickly integrate DAB audio and data capability into their in-car systems.

This paper will cover the following points:
- Advantages and applications of DAB in the automotive marketplace.
- The design challenges inherent in implementing DAB in an automotive environment.
- Example automotive DAB solutions.
- Future roadmap for digital radio.

1   Advantages of DAB in the Automotive Environment

Eureka-147 enjoys a number of significant benefits to the automotive listener, and also some significant benefits to the broadcasters. With the DAB service being heavily promoted by broadcasters in Europe the combination of these benefits and the new content that is being broadcast is making DAB a compelling feature for cars that will be manufactured in the coming years.

2   Benefits of DAB

High quality digital audio with rugged and reliable reception – listeners can benefit from a good quality of reception in a moving environment through the immunity to interference that is inherent in the DAB system. Coupling this with the excellent audio quality that high bit rate digital audio provides ensures that listeners soon start to prefer their digital radio over FM.

Broader range of audio services – as DAB offers more efficient spectrum usage than analogue FM a key benefit to the listener is the availability of more and diverse services. For example, across the UK there are now over 100 different radio channels (national and regional) available to the listener. In the case of the BBC, DAB has enabled the broadcaster to make available eleven high quality audio channels over the DAB coverage area (currently 65% of the UK population, rising to 85% in 2004), compared to the four national channels available over FM.

Easy to use and informative receivers – tuning can be easily be performed via a text based user interface. The receiver can supply to the listener a comprehensive list of services that are available, a list that can be simply navigated via the user interface. This is coupled with the ability of receivers to follow a selected service without the need to re-tune.

More efficient use of broadcast spectrum, through the audio compression technologies utilised in DAB broadcasters can achieve significant gains in the number of audio services that can be transmitted with good quality in a given amount of bandwidth. As has been noted above this feature has already been used by broadcasters in the UK to enable them to provide a larger number of services than they are able to over FM.

3   Data Services – Benefits to Users and Broadcasters

DAB can be considered as a fast and low cost data pipe. Approximately 20% of DAB bandwidth is reserved for data applications, which are split into three areas as shown in the table below.

There is increasing interest in utilising the DAB to connect drivers with more useful information such as navigation updates and more detailed traffic information via Traffic Message Channel (TMC) and Traffic Program Expert Group (TPEG) standards. These services can be delivered free to air, or using a cellular connection to provide a back channel that will enable service providers to charge for more valuable services.

Service Name	Service Description
Service Information (SI)	Basic RDS like functions to underpin the service
Program Associated Data (PAD)	Used to broadcast information about the station being received and the content being played
Non-program Associated Data (N-PAD)	Available for other applications such as map data, dynamic navigation updates etc.

Table 1.   DAB data applications.

The data capacity of DAB is increasing in importance with the advent of dynamic navigation systems where the quantity of data required to update the static maps with traffic news will start to exceed the capacity of the FM RDS channel.

4   Design Challenges

One of the main characteristics of DAB is the complexity of the air interface. This complexity has been driven by the requirement to provide excellent audio quality even in poor signal conditions. The signal chain implemented in the base band processor requires a great deal of real time computing power, and also the interleave depth specified in DAB requires a large quantity of memory.

Coupling these processing requirements with the other environmental challenges faced in an automotive environment (temperature, vibration, limited ventilation and extreme cost pressure) makes implementing a cost effective automotive DAB system a difficult problem to address.

5 Processing Trade Offs

As is noted above, DAB requires a significant amount of processing power. This processing power can be provided by a fast general purpose DSP, and examples in the market show that a 200 MHz dual MAC DSP can provide sufficient horsepower to implement DAB.

However, it is also worth noting that general purpose DSPs driven at high speed also dissipate a lot of power (an example is a DSP that is being put forward as a DAB engine, will dissipate 280 mW at the required 200 MHz), which could bring ventilation problems in a crowded modern head unit.

Because of the power limitations that are inherent in a general purpose DSP, another approach is to implement the DAB signal processing chain in dedicated hardware. This approach has the benefit of reducing power consumption (generally gates will consume less power than a software implementation), but this power consumption saving comes at the expense of system flexibility. In the DSP approach, spare processing power can be used to implement other radio features such as a UI or command interface, whereas in a hardware solution this option does not exist and a host processor is required even in the simplest applications driving up system cost.

A third approach attempts to marry the DSP and hardware solution to achieve a good compromise of flexibility, cost and performance. This approach has been taken in the Frontier Silicon Chorus I multimedia processor, and it has resulted in a low power (and low cost) DAB base band solution that can also perform other tasks such as MP3 decode.

6 Cost Pressure

In the consumer market, stand alone DAB receivers are now being sold for under €100. Especially in the UK, this has helped the volumes of DAB products take off very rapidly; with some estimates showing that over 500 K receivers were sold in the UK during 2003.

Cost has been driven down by companies like Frontier Silicon deploying highly integrated base band devices for DAB that significantly reduce the bill of materials required to implement a radio, and also providing high performance and low cost modules that enable consumer manufacturers to easily implement DAB in their products.

The low component count approach that can now be achieved in a DAB system is demonstrated in the diagram below showing a base band implementation for a Frontier Silicon base band module implemented using just three ICs, of which two are commodity serial flash and stereo codec devices.

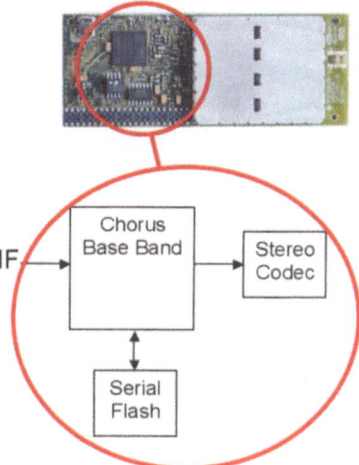

Fig. 1.   Low chip count DAB base band.

7   Frontier Silicon DAB Solutions

To implement high performance DAB solutions, Frontier Silicon has teamed up with several key players. These included Ensigma (a division of Imagination Technologies), TSMC, Flextronics and The Technology Partnership.

Based on the Frontier Silicon Chorus I (FS1010) DAB base band, this team has developed and produced a range of DAB modules to meet a number of different market requirements. The basic features of these modules are described below.

Chorus I is based on a multi-threaded, dual MAC DSP core operating at 150 MHz, and as it is targeted a DAB applications it includes a number of "DAB accelerators" for computing intensive parts of the DAB system such as the Viterbi. To enable fast and low power execution of the various algorithms required in a DAB system FS1010 also includes 384 Kbytes of SRAM. This means that the base band can be implemented with just one, low cost, serial flash memory for program and parameter storage.

The DAB hardware and software in the Chorus I enable the device to perform the demodulation, decoding, protocol stack required to decode DAB using just 30% of the processing bandwidth available when the device is operating at 150 MHz, giving a full DAB base band solution that consumes less than 100 W.

Module Name	Target Applications	Features
Venice (1.1 and 2.0)	Kitchen Radios, CD boom boxes, Micros, Tuners, Clock Radios High End Hi Fi All Mains powered audio	Dual Band (Band III and L Band) reception, stereo audio out, Digital Audio Out, LCD controller, Keypad interface Slave DAB module mode
Diablo	Hand Held/ Pocket Radios Portable radios, Jukeboxes	Dual Band (Band III and L Band) reception, Very small size, low power consumption, single power supply input, stereo and digital audio out, LCD controller, Keypad interface
Roadster	Integrated car radio/CD players, navigation systems, telematics	Dual Band (Band III and L Band) reception, small size low power consumption, digital audio out, automotive software extensions

Table 2.

8   Block Diagram of Chorus FS1010 Media Processor

This intelligent combination of hardware and software has resulted in a very low power solution that maintains the software flexibility that is characteristic of DSP based solution.

As well as a powerful DSP and DAB accelerators, Chorus I is also a very highly integrated system on a chip (SoC) device. It integrates a large number of the peripherals that may be used in a DAB receiver, including a USB interface and LCD controller. The aim of this integration is to enable systems to be developed with DAB capability, but with maximum flexibility to enable the base band to fulfil a number of other system requirements, so reducing manufacturing cost.

For example, the Venice and Diablo modules that are based on Chorus I are capable of performing the user interface function as well as performing DAB decode, therefore removing the need for a further microcontroller and associated memory and peripherals in the system.

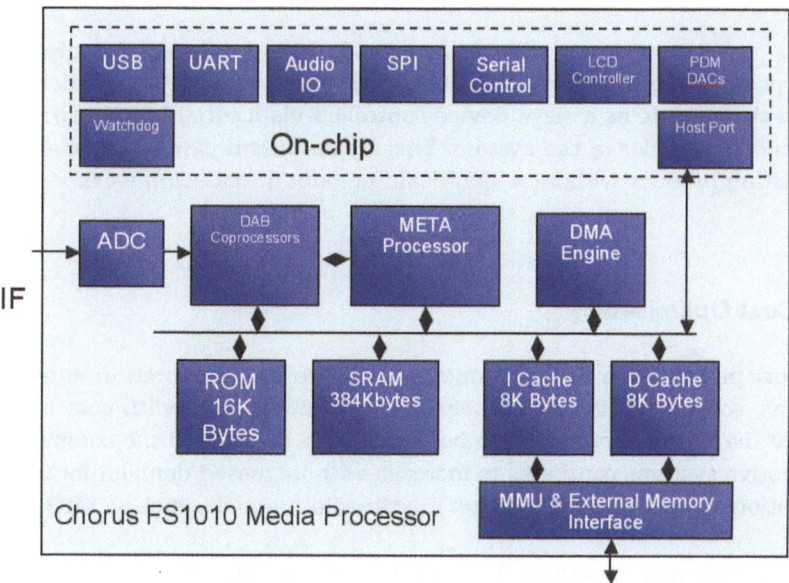

Fig. 2.   Block Diagram of chorus FS1010 media processor.

9   Roadster Automotive DAB Module

A real world example of an automotive DAB solution is the Frontier Silicon Roadster module, which has been designed in partnership between Frontier Silicon and The Technology Partnership.

The design objectives of Roadster were generated from intensive discussions with customers to make the specification as close as possible to the requirements of customers.

The key requirements that prospective customers put forward were:
▶ Good sensitivity and general RF performance.
▶ The module should be small enough to fit into a 1 DIN integrated car radio/CD player.

▶ Power consumption should be as low as possible.
▶ A small number of connections to the main pcb.
▶ Controlled as a slave module via a serial interface.
▶ Software features such as DAB to FM service linking to support the automotive environment.
▶ Ability to perform other multimedia tasks such as MP3 decode.

As the Roadster module will be installed in many different integrated car audio product (e.g. a combined DAB + AM/FM Radio with CD) it was decided to design the module as a slave device controlled via a serial interface from the host microcontroller in the system. This means that it can be easily added to an existing product without a significant amount of re-design work.

10 Cost Optimisation

The cost pressures in the consumer market are also mirrored in automotive systems, so automotive DAB receivers must be designed with cost in mind, despite the heavy complexity embedded within them. And the complexity of automotive systems continues to increase with increased demand for low cost navigation systems, as well as new multimedia functions such as MP3.

A key opportunity to reduce the cost of implementing DAB in an automotive system is through integration. Many 1st and 2nd generation DAB receivers on the market are implemented as "black boxes", connected to the audio head unit via screened cables and controlled via a CAN interface or via a multimedia bus such as D2B or MOST.

This approach may be valid where it is not feasible to implement DAB alongside the other functions in the audio head unit. However, the "black box" approach has added expenses – the module implementing the DAB function duplicates several functions in the head unit, including microcontroller/DSP, power supplies, network interfaces and connectors.

A more cost effective approach requires a DAB function that can be implemented as part of a multi-function audio head unit. This requires that the DAB module is small enough (a good benchmark is the size of an existing AM/FM tuner) and also does not consume a great deal of power to avoid internal heating effects and to minimize the effect on the power supply design.

As can be seen from the block diagram, the Roadster module has been designed with minimal connections, enabling a simple interface to the rest of the system as well as keeping the size of the module to an absolute minimum. The dimensions of Roadster are just 55 mm long by 37 mm high by 8mm wide, making it competitive in terms of size with most AM/FM tuners. Power consumption is also very low, with the full receiver consuming just 800 mW while decoding a DAB signal and outputting audio.

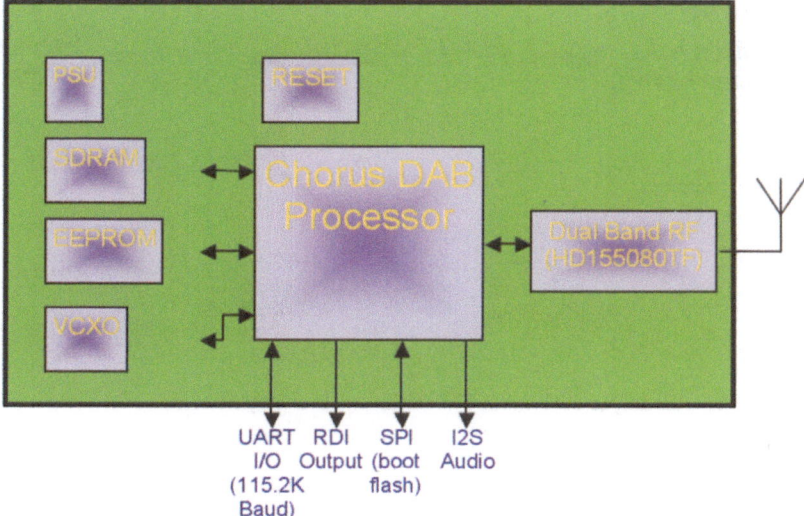

Fig. 3.   Roadster block diagram.

The size and power consumption of Roadster make integration of the DAB function into a crowded radio/CD/MP3 1DIN device.

As well as the standard DAB demodulator, protocol stack and master/slave interface that is found in the consumer oriented modules from Frontier Silicon the Roadster module also takes advantage of the extra information broadcast to support features such as service linking. Improved soft error concealment strategies are also included to improve the listening experience in poor coverage areas.

11   Example DAB+AM/FM Radio + CD Using Roadster

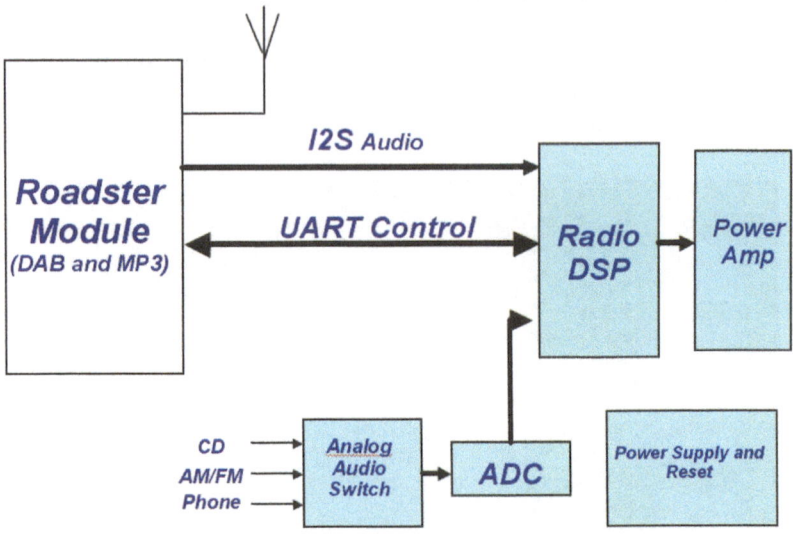

Fig. 4.   Example DAB+AM/FM radio + CD Using Roadster.

12   DAB Roadmap

The demand for DAB receivers is already taking off rapidly, especially in the UK, as it is expected to grow rapidly in other European countries as new services become available solely over the digital broadcast medium. In the automotive market the drive for equipping cars with DAB will come from two directions, from car purchasers who want to enjoy the new audio services and clearer reception, and from car purchasers who wish to have the latest navigation technology.

In both cases, the cost of installing DAB in the car is going to be crtical – it has already been noted in this paper that one of the catalysts for DAB in the UK has been the availability of receivers costing less the € 100, and this factor will also influence car buyers and also the OEMs who make the decision to integrate the functionality in their new vehicles.

To meet the increased cost pressure semiconductor manufacturers will have to respond with further improvements in their DAB solutions, concentrating on reducing the silicon real estate required to implement the DAB function.

As DAB matures there is also an increasing trend to use it as a transport medium for data in other applications. A very real example is the decision of the Korean government to introduce a multimedia broadcast service in 2004 called DMB (Digital Multimedia Broadcast) that builds on the Eureka 147 specification.

Fig. 5.   Frontier Silicon Roadster on Evaluation Board.

DAB has taken a long time to arrive, but with the convergence of new audio services being broadcast solely over DAB and low cost, consumer friendly receivers becoming available it is clear that the time for DAB has come.

Steve Evans
Frontier Silicon Limited
Innovation Centre
Home Park Estate
Kings Langley
Hertfordshire, WD4 8LZ
United Kingdom
steve.evans@frontier-silicon.com

Automatic Functional System Test of Complex Automotive Devices

A. Guerra, IXFIN Magneti Marelli Sistemi Elettronici

Abstract

The purpose of this paper is to show how starting from the customer specifications the basic methodologies of software development have been applied to the realizations of two ATE (Automatic Testing Equipments) for the functional verification and validation of two automotive devices with specific complex constraints in the functional and real time domains, with body/comfort and HMI functionalities.

The problems of formal accuracy and exactness of notation present in the software development cycle that may affect the correct functional and real time behaviour of the resulting target system will be analyzed. It is then described which were the general purposes and features identified as necessary for the construction of the two ATE's, that led to the definition of the hardware and software solutions. Finally, the technical characteristics of the machines resulting from this study are described in two separate paragraphs, were is evident how devices from different domain led to different solutions.

1   Introduction

The latest generations of automotive electronic devices are introducing more and more complex functionalities, with more demands on software and hardware performances. From the combination of functionalities like voice recognition, graphic elaboration, and mpeg decodification in the infotelematic domain, to the hard real time requirements necessary for safety related systems.

This complexity is obviously reflected in the whole process of the system development till its validation. Starting form the customer requirements specifications the software life cycle passes through several phases, each one of whom can introduce indetermination in the final results. Nowadays this process in the automotive applications still suffers of several problems that can influence in a bad way these indeterminations. It's compulsory therefore the support of precise methodological foundations. The system validation of the device is the ultimate reverse direct link with the initial customer requirements, therefore if performed with a rigorous approach is the consistent confirmation of the effective compliance of the results with the initial design. Moreover, the system validation is gaining more importance as the software and the hardware are becoming more and more a unique component, and their associated performance is becoming sensibly visible and important for the final user.

Due to this increasing complexity the system validation performed with usual non-automated tools is time consuming and not adequately efficient. Combined functionalities with real time demands specifications require long time complex testing activities. Elaborated HMI functionalities are hard to measure and validate (e.g. the timing measurements of system response on a display panel to different contemporary stimuli coming from the operator and field signals). The coverage analysis and insight of the functional domain is evidently an issue if the system has intricate functions. Furthermore the non-automated methods do not allow the correct repeatability of long system "on the field" tests.

This considerations convinced Magneti Marelli's R&D testing department that a specific ATE (Automatic Testing Equipment) with a rigorous real-time and behavioural engine able to perform complex and repetitive tests would solve this problems, improving the efficiency of systems testing and validation. An ATE capable of generating test sequences starting from a formal high-level description of the system specifications with functional and real time information, and endowed with the appropriate simulators and actuators able to fully emulate the operational field of the system was therefore needed.

2   The Problems in the Software Life Cycle

During the software development cycle there are many formal accuracy problems that we can encounter that affect the correct functional and real time behaviour of the resulting target system. Figure 1 shows for reference a common V-Model representation of the various stages in the software development.

First of all at the present time and for almost every system design, the requirements received from the customer are usually expressed in an informal notation, with a large use of natural language, and are comprehensive of functional and non functional behaviour rules, real time aspects, till to the definition of
EMC and lifetime issues. We usually start therefore from a system's design
basis that is lacking of a formal description. A direct consequence is a certain
lack of precision in the description, that is, even if rigorous, quite difficult to
manage from a technical point of view.

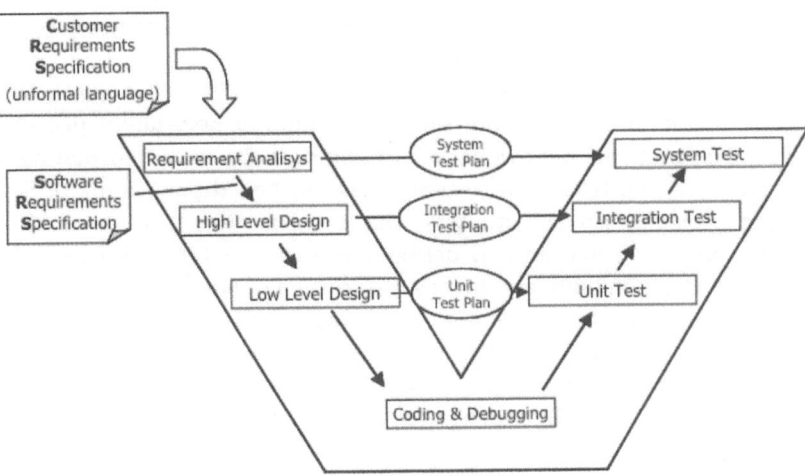

Fig. 1.   The software life cycle.

Proceeding in the software lifecycle the customer requirements specifications
are collected, analyzed, discussed and reviewed with the customer itself in
order to translate them into a consistent software requirements specification.
This is a first step of conversion that introduces to the problem of "fragmentation". With "fragmentation" [7] is intended that the link among the different
phases does not occur through a direct scientific and rigorous conversion but
may require manipulation and intervention. Therefore the flow of information
through the different stages is fragmented. Of course this can affect the precision of the implementation.

The software requirements specifications are then the input for the subsequent design development. At the high level design the overall software architecture with the relationships among the required software components is
defined. Real time constraints and performances have to be correctly
addressed, in consideration of the lower hardware platform. Hardware and
software partitioning are important decisions at these early stages. In the

design procedure, leading from the high level to the low level design, arises the problem that not all the software and hardware performances are well known. It's impossible to foresee the real time behaviour of software modules when executed on the field by the microprocessor, therefore hypothesis on the average performance of those modules are used.

To solve these problems there are many powerful tools that can provide a pseudo-coding starting from a high level design. These software development environments are specifically designed to describe and write real time code that runs on embedded systems. In any case, the initial lack of rigorous translation of the specifications still exists each time the customers give to the supplier an informal requirements description.

Different studies are carried out on the pursuits of new languages able to describe the functional behaviour and timing description in a formal notation as, for example, the UML, that recently has reached the version 2.0. Several European projects also are studying the problem. Not last, the EAST-EEA European project is working to define the ADL, Architectural Description Language, a common language for the description of automotive architectures and components and the formats to exchange description data between suppliers and integrators. The ADL is also based on UML standards, and its approach separates the automotive electronic architecture in multiple layers, from a top vehicle view till to an abstraction description of the lower hardware layer.

Once the software development and testing have a positive result, the following phase, which is the main purpose of this paper, is the realization of the System Testing.

2   Realization of the System Test

In the following two paragraphs, it is shown how two specific ATE, one for a body control device and one for a telematic module, have been first defined and then developed, starting form the customer specification requirements. The analysis of what the system is and how it operates in its working environment has been the starting point of the work. It must be underlined an important characteristic of the system testing: the system test must be performed with a black box approach, i.e. without opening the system, and running the end-user software. The rationale to this approach is that any invasive technique used to extract information from the system itself can potentially perturbate the measures, and therefore must be avoided. All the interactions have to be performed using the system user interfaces, in the field environment or a simulation of it. In other words the system has to remain as it will be sold on the market [1]. An intensive analysis on the system's input and output devices has been therefore carried out to create a perfect reproduction of the real stimuli with dedicated simulators (keyboards, audio inputs and outputs, CAN Network, simulation of the GSM network RF signal and GPS RF signal, etc.). For all these it has been analysed how to implement on hardware and how to pilot their interaction with the system under test.

Another important issue is the real time domain management. Automotive applications have strict real time requirements, which therefore become an important feature for the system under test as well for the simulated environment to which it is connected (the ATE).

Beyond the hardware of the ATE there is its core engine that has the purpose to manage the testing activities in a real time domain. The main general functions identified for the ATE are:
- ▶ Simulate the environment around the connections of the device under test through an hard real time, event driven control of the simulators.
- ▶ Execute automatically the test sequences that simulate the identified necessary operative conditions.
- ▶ Acquire and record all the subsequent conditions/states.

With the objective to:
- ▶ Verify the application software in the time and value domain.
- ▶ Verify the robustness of the software.
- ▶ Manage the errors.
- ▶ Get evidence of the criticital functionalities.

ATE's runtime management strategies and programming methodologies have been object of studies in collaboration with the technical university "Politecnico di Torino" with the intent to obtain an ATE capable of generating test sequences starting from a formal high-level description of the system with functional and real time information. This would be the way to create the strong link with the initial specification that if performed with a rigorous approach would represent the consistent confirmation of the effective compliance of the results with the initial design. The language chosen as standard notation is UML. This language is also a valid way to describe timing attributes. In fact there's the possibility to describe absolute and relative actions timings referred to the sequence diagrams. The input for the ATE would come from the same UML high-level description used to describe the scenarios of the system behaviour. This would save time and avoid the fragmentation problem described before.

3   The Realization of the Telematics Domain ATE

Ixfin Magneti Marelli Electronic Systems designs and produces advanced electronic systems and modules for vehicles. The infotelematic system subject of the test is a system that delivers to the user services and options like GSM, Radio, voice recognition and speech, Audio CD, WAP, MP3 and Navigation with GPS. A very important characteristic of any infotelematic system is the HMI (Human Machine Interface), which allows the user to interact with the system trough the video and a set of keys and/or knobs.

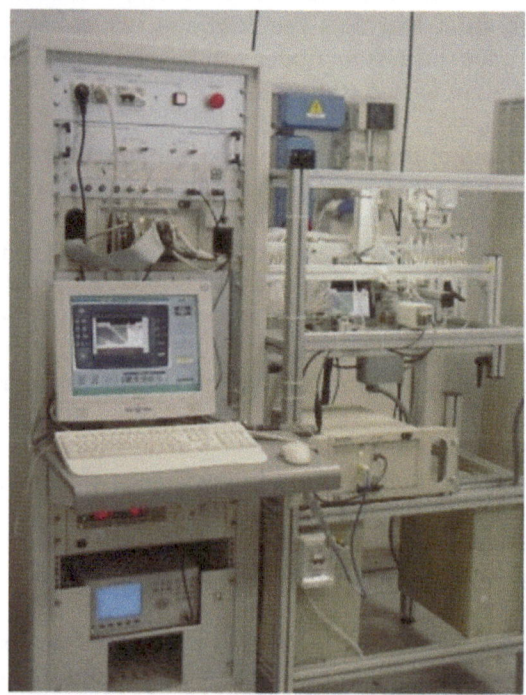

Fig. 2.   Telematic ATE.

The software verification is a part of the development process that is very important on a system with this complexity. With these level of HW and SW a lot of functions are working at the same time (multitasking environment) and it is very difficult to simulate the environment interaction, i.e. the operations performed by the user, the car network, GPS end GSM signals, etc.

The solution:

Ixfin Magneti Marelli R&D Testing department has developed an Automatic Test Equipment (ATE) based on a "National Instrument" PXI computer instrumentation with vision, motion, speech synthesizer, actuators and special test sets, controlled by LabView and test stand environment. More in details the test equipment is based on two parts:

▶ The main PXI System, including: the CAN interface, the audio card for the speech synthesizer and the voice recognition; an image acquisition board; a digital I/O board used to control the internal matrix of the main PXI System; a multifunctional I/O device used to generate or acquire digital signals; a controller of the instrumentation on the main PXI. The other instrumentation used is the GSM network simulator, the radio broadcasting synthesizer simulator and the RDS generation. As a GPS system a dedicated device directly connected on the specific connector of the infotainment module and able to simulate the RF satellite signal, that would be received during some customizable trips, is used.

▶ The slave PXI, more related to the realization of the HMI interaction which consists in an automatic mechanical actuator for the keys pressing, controlled trough a PLC managed by a PC serial line, a colour camera for the verification of the Display, and two devices for the automatic insertion/ejection of the CD and of the SIM card for the telephone.

With this structure it is possible to perform robustness tests simulating the use of the product in the car, running at the same time sequences of emulation of interaction with the user (HMI), during parallel processing of radio signal, RDS, GSM communication, GPS reception, CD-Player, data management, and communication with the car trough the CAN interface.

4 The Realization of the Body Domain ATE

This has been the realization of an ATE to test devices for body/comfort applications, i.e., internal and external lights, commands, serial interfaces, comprehending 2 CAN network interfacing with gateway functionalities. This system required more rela time constraints that influenced the choice of the operative system running on the ATE. First the customer specifications of the device have been analysed. The analysis has produced the specifications for the system ATE, divided in hardware and software. In the hardware specifications have been described the requested features:

▶ Power management of the device under test.
▶ Analogical, logic and frequency signal generation.

▶ Fault simulation on every contact of the device, i.e. short circuit to ground and to battery, and disconnection.

▶ Electrical characteristics of the device under test as well for the expected loads connected.

The software specifications required:

▶ The possibility to create and modify test sequences.

▶ The possibility to graphically monitor the stimuli produced by the ATE and the output generated by the system.

▶ The possibility to manage and configure all the CAN network messages.

▶ Generated with timing relations or event based.

▶ Reporting.

Fig. 3.   Body ATE.

The Solution:

The hardware of the system obtained is composed of a PXI rack with the boards dedicated to controller, serial communication ports (CAN, RS232), signal matrices, power matrices, resistor matrices, I/O to GND/Vbat TTL conditioners, relays, characteristic impedances, etc. The electrical connection of the device under test is obtained with a specific fixture.

The software is composed of four combined tools that allow the following testing procedure: to generate on a host PC a test pattern (pattern editor), download it on the real-time controller of ATE (RT controller manager), execute it with the report generation of the events (real time program), download it on the Host PC (RT controller manager), and analyze it to check the behaviour of the device (data analyzer). In particular: the pattern editor allows and assists the user in the definition of a testing sequence, using functions databases and mnemonic names for the system resources; the data analyzer allows to analyze the values of the events and their correlations in the time domain, also showing the data passed on the communications networks.

5 Conclusions

With the use of the testing equipment and methodologies described above, Ixfin Magneti Marelli Electronic Systems gained many advantages in terms of reliability, time-to-market, and development costs. Our software robustness tests assure that the product in output is more reliable; the time for verification is reduced and it's not necessary to dispose of a real prototype car for the testing. Complex, repeatable, specifically aimed tests that a normal human user could not perform, with this equipment can be. This helped for example to find out anomalies and problems that were occurring only after several cycles, and that a manual testing could not point out. Moreover through the automated testing it has become easier to guarantee the functional reliability and no regression capability on previously developed software parts.

From the methodological point of view, is obvious that is becoming more and more necessary a standardization of the languages and the formats for exchange of description data between suppliers and integrators that would definitively set up the strong rigorous approach for software developments.

6 Acknowledgements

I would like to thanks all the persons who directly got involved in the studies hereby described and who helped the realization of this article: Mortara P., Mo S., De la Pierre P., from Ixfin Magneti Marelli Electronic Systems; and Baldini, A., Benso, A., Prinetto P. from the Politecnico di Torino.

References

[1]   Beizer, B. – "Black-Box Testing: Techniques for Functional Testing of Software and Systems" – New York – John Wiley and Sons – 1995

[2]   Bradfield, J. C. – Verifying Temporal Properties of Systems (Progress in Theoretical Computer Science in Theoretical Co) – Springer Verlag – January 1992

[3]   Gomaa, H. – "Sofware Development of Real Time Systems" – Communications ACM 29 – no.7 – page(s) 657-668 – July 1986

[4]   Jin Z., Offutt J., Abdurazik A. and White E. L. – Analyzing Software Architecture Descriptions to Generate System- level Tests – Workshop on Evaluating Software Architectural Solutions 2000 (WESAS) – May 2000

[5]   Fernandes, J.M.; Machado, R.J.; Santos, H.D. – Modelling industrial embedded systems with UML – Hardware/Software Codesign, 2000. CODES 2000. Proceedings of the Eighth International Workshop on page(s): 18 – 22 - May 2000

[6]   Douglass, B.P. – Real Time UML – Addison Wesley Pub. – Oct. 1999

[7]   Baldini, A., Benso, A., Prinetto, P., Mo, S., Taddei, A. - Towards a Unified Test Process: from UML to End-of-Line Functional Test - International Test Conference 2001 (ITC'01) - Proceedings - page(s) 600-608 - Oct. 2001

Andrea Guerra
IXFIN Magneti Marelli Sistemi Elettronici
Viale Carlo Emanuele II, 118
10078 Venaria Reale (TO)
Italy
andrea.guerra@ixfin-mmarellise.com

Keywords: Automatic Testing Equipment, HMI Body/Comfort Devices, System Test, Functional Verification

Telematics Digital Convergence – How to Cope with Emerging Standards and Protocols

K. Parnell, Xilinx

Abstract

Digital convergence, in recent history, has been prevalent in the consumer equipment domain and the design engineers in this area have been struggling with a plethora of emerging standards and protocols. What lessons can we learn from their struggle? The same dilemmas now exist in in-vehicle telematics and infotainment systems but with the added issues of extremes of temperature, safety, security and time in market.

This paper will first describe and position each emerging in-vehicle standard and their respective strengths and weaknesses. It will then explore the designers dilemma: how to build flexible and scalable system architectures which will allow the time in market of telematics platforms match that of the host vehicle whilst still communicating internally and to other external systems. It will then go on to discuss enabling technologies and how to implement reconfigurable and upgradeable telematics platforms that can cater for protocols today and in the future.

1   Introduction

In-car electronics content has seen tremendous growth in recent years, not only in the traditional body control and engine management areas but in the new areas of driver assistance systems and telematics type applications. Figures recently published by the IEEE show rises in car electronics of up to 16% and they announced that by 2005 electronics will account for 25% of the cost of a mid-size car.

One of the high growth area is telematics systems (the convergence of mobile telecommunications and information processing in cars) which exhibit characteristics more like those of consumer products i.e. short time to market, short time in market and changing standards and protocols. These characteristics are the complete opposite to those of traditional in-car electronics, which could potentially cause issues from design right through to manufacture and servicing. These issues have an impact on the way designers are approaching designs and also which in-car bussing system to choose.

The traditional in-car bus systems based on a serial, event triggered protocol such as CAN and J1850 have been used in the body control area successfully for many years but bandwidth and speed restrictions will make it difficult for these to be used in the newer real time applications. A range of new bus standards have been emerging such as time triggered protocols and optical data busses to meet these new data through put challenges. In-car bus networks can be divided into four main categories:

- ▶ Body Control – covering data and control signals between car seat controls, dashboard/instrument panel clusters, mirrors, seat belts, door lock and airbags (passive safety).
- ▶ Entertainment and Driver Information Systems – communication and control between radio, web browser, CD/DVD player, telematics and infotainment systems.
- ▶ Under the hood – networking between ABS brakes, emission control, power train and transmission.
- ▶ Advanced Safety Systems – data transfer between brake-by-wire, steer-by-wire and driver assistance systems (active safety).

We will look at each area in turn and discuss the different emerging vehicle emerging standards and protocols that apply to each area. Figure 1 and figure 2 show the various in-car networks and their intended applications.

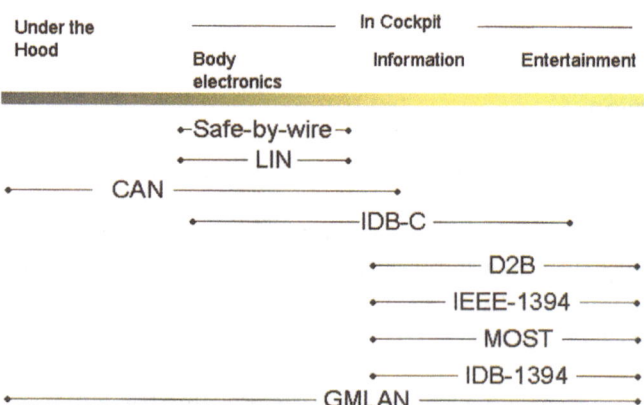

Fig. 1.   Applications for in-vehicle networking.

2   Body Control and 'Under-the-Hood' Busses

2.1   Controller Area Network (CAN)

In the mid-1980s Bosch developed the Controller Area Network (CAN), one of the first and most enduring automotive control networks. CAN is currently the most widely used vehicle network with more than 100 million CAN nodes installed. A typical vehicle can contain two or three separate CANs operating at different transmission rates. A low speed CAN running at less than 125 Kbps for managing the body control electronics such as seat and window movement controls and other simple user interfaces. Low speed CANs have an energy saving sleep mode in which nodes stop their oscillators until a CAN message wakes them. Sleep mode prevents the battery from running down when the ignition is turned off.

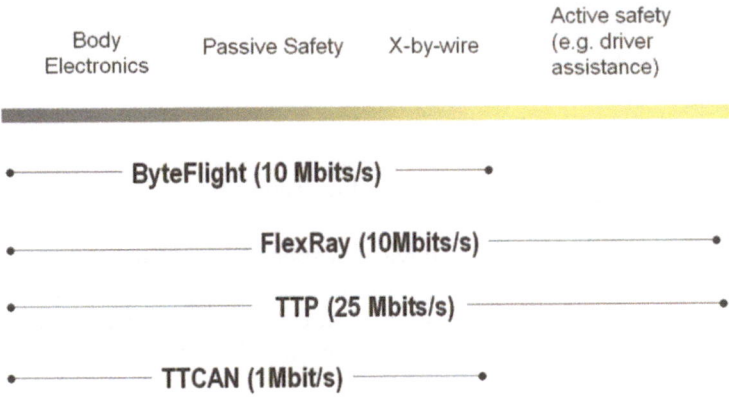

Fig. 2.   Safety Critical Busses.

A higher speed (up to 1 Mbps) CAN runs more real-time-critical functions such as engine management, antilock brakes and cruise control. Controller-area network (CAN) protocols are becoming standard for under-the-hood connectivity in cars, trucks and off-road vehicles. One of the outstanding features of the CAN protocol is its high transmission reliability, which makes it well suited for this type of application. The CAN physical layer is required for connections to under-the-hood devices and in-car devices.CAN has found its way into general industrial control applications such as building management systems and lift controllers but more surprisingly is now being considered for use in telecoms type applications such as Base Transceiver Stations (BTS) and Mobile Switching Centers (MSC) equipment. This is due to CANs protocol error management, fault isolation and fault tolerance capabilities.

2.2 Local Interconnect Network (LIN)

The Local Interconnect Network (LIN) has been developed to supplement CAN in applications where cost is critical and data transfer requirements are low. Conceived in 1998, the LIN consortium comprises car manufacturers Audi, BMW, DaimlerChrysler, Volvo and VW. LIN is an inexpensive serial bus used for distributed body control electronic systems in vehicles. It enables effective communication for smart sensors and actuators where the bandwidth and versatility of CAN is not required. Typical applications are door control (window lift, lock and mirror control), seats, climate regulation, lighting and rain sensors. Outside the automotive sector LIN is used for machine control as a sub-bus fro CAN.

LIN is a UART-based, single-master, multiple-slave networking architecture originally developed for automotive sensor and actuator networking applications. The LIN master node connects the LIN network with higher-level networks, like CAN, extending the benefits of networking all the way to the individual sensors and actuators.

3   Entertainment and Driver Information Systems

Car infotainment and telematics devices, especially car navigation systems, require highly functional operating systems and connectivity. Until now the existence of both open-standard and proprietary stand-alone buses has worked quite well. But because of their integrated nature, future systems will require electronic subsystems to work together.

By relying on open industry standards, all key players – from manufacturers to service centers and retailers – can focus on delivering core expertise to the end user, rather than expending the time and effort it would take to develop separate, incompatible designs for specific vehicles or proprietary computing platforms. Future systems will be highly integrated, open and configurable.

Several organizations and consortia are leading the standardization efforts, including the MOST Cooperation, the IDB Forum and the Bluetooth Special Interest Group (SIG).

3.1 Media Oriented System Transport (MOST)

MOST networks, managed by the MOST Cooperation, are used for connecting multiple devices in the car including car navigation, digital radios, displays, cellular phones and DVDs. MOST technology is optimised for use with plastic optical fibre, supports data rates of up to 24.8 MBit/s, up to 24 nodes, infinite length, is highly reliable and scalable at the device level, and offers full support for real-time audio and compressed video. MOST technologies are strongly supported by German automakers and suppliers. The MOST bus is endorsed by BMW, Daimler Chrysler, Harman/Becker and OASIS Silicon Systems. A recent notable example of MOST implementation was its use by Harman Becker in the latest BMW 7 series. Optical data buses in cars have recently been favoured for their noise immunity characteristics but there is concern that repeated connection and re-connection of the optical connectors has been found to cause damage to the optical fibre (typically 5-6 re-connects). Optical fibre fractures and damage is notoriously difficult to find so some automotive companies are reviewing MOST with a view to maybe moving back to copper wire instead of optical fibre.

3.2 Intelligent Transport System Data Bus (IDB) 1394 and CAN

The IDB Forum manages the IDB-C and IDB-1394 buses and standard IDB interfaces for OEMs for the development of after-market and portable devices. Based on the CAN bus, IDB-C is geared toward devices with data rates of 250 kbits/s with up to 16 nodes. Applications for IDB-C include connectivity through consumer devices such as digital phones, PDAs and audio systems.

IDB-1394 (based on IEEE-1394 Firewire) is designed for high-speed multimedia applications that require large amounts of information to be moved quickly on a vehicle. IDB-1394 is a 400 MBaud network using fibre optic technology with a maximum of 16 nodes over 72 meters. Such applications include DVD and CD changers, displays, and audio and video systems. IDB-1394 also allows 1394 portable consumer electronic devices to connect and interoperate with an in-vehicle network. Zayante is one supplier who currently has 194 physical layer devices for the consumer market and demonstrated IEEE-1394 with the Ford Motor Company. The demonstration included plug and play connections of a digital video camera, Sony Playstation II as well as two video displays and a DVD player.

Already a leading standard in consumer electronics, 1394 is now almost ready for prime time in vehicles. It's ability to simultaneously stream multiple channel of audio and high-quality video make it the technology that can transform vehicles into the multi-screened rolling entertainment centres envisioned in many concept cars. And work is underway to make it practical for the demanding automotive environment.

Some still think that 1394 is "over kill" for the automobile whereas others view it as the next stage after MOST. There is also a view that MOST and 1394 will coexist, MOST for the embedded infotainment applications and 1394 for high speed, real time video applications. The interest for 1394 is growing in Japan and Europe but North American manufacturers still appear to be neutral.

1394 should start showing up in production vehicles in 2006, but may appear in after market products sooner. Renault are publicly leading the charge for 1394 adoption.

3.3 The Digital Data Bus (D2B) Communications Network

The Digital Data Bus (D2B) is a networking protocol for multimedia data communication integrating digital audio, video and other high data rate synchronous or asynchronous signals, can run up to 11.2 Mbps, and can have up to 50 nodes (copper) and 24 nodes (optical). The data bus can be built around either unshielded twisted pair (UTP), named 'SMARTwire' or a single optical fibre. The maximum bus length for copper is 150 meters and no limit to the optical variant. This communication network is being driven by C&C Electronics in the UK and has seen industry acceptance from Jaguar and Mercedes Benz. For example the Integrated Multimedia Communication System that has been deployed in the Jaguar X-Type, S-Type and new XJ saloon.

D2B optical is self configuring on start up, adapting to what ever devices are present on the network at the time. This means that new devices can easily be fitted to the network at any time during its life. Car makers that use the D2B optical multimedia system will find the standard evolves in line with new technologies as they are introduced. The standard will be backwards compatible ensuring that new products can be added to a cars system during its life time. D2B optical is based on an open architecture which simplifies expansion as no changes to the cable harness are required when adding a new device or function to the optical ring. The buss uses just one cable, either a polymer optical fibre or copper to handle the in-car multimedia data and control information. This gives better reliability, fewer external components and connectors and a significant reduction in overall system weight.

3.4 Bluetooth

Bluetooth wireless technology is a low-cost, low-power, short-range radio technology for mobile devices and for WAN/LAN access points. It is a computing and telecommunications industry specification that describes how mobile phones, computers and PDAs can easily interconnect with each other, and with home and business phones, and computers using a short-range wireless connection. Bluetooth is an open wireless standard that uses low-power, short-range (about 35 ft.) radio signals at the unlicensed 2.4 GHz frequency. It has a spread-spectrum, full-duplex signal that hops frequencies at up to 1600 times/sec to reduce interference, and a data transfer rate of about 1 Mbps.

A driver will be able to use a Bluetooth cordless headset to communicate with a cellular phone in his or her pocket, for example. As a result, driver distraction can be reduced and safety increased. The automotive industry has created a special-interest group (SIG) for the definition of Bluetooth car profiles. The SIG includes such members as AMIC, BMW, DaimlerChrysler, Ford, GM, Toyota and VW. One example of Bluetooth deployment in cars is Johnson Controls BlueConnect™ which is a hands-free system that allows drivers to keep their hands on the wheel while staying connected through a Bluetooth-enabled cellular phone.

There has however been some concern voiced over the long term support of Bluetooth devices and how noisy in-car environment will effect its operation. The lifecycle of cars and other vehicles is very much longer than that for consumer products or mobile phones so this mismatch between support and service timescales must be addressed by silicon manufacturers. However the Chrysler Group showed the use of Bluetooth connectivity in its vehicles at Convergence 2002.

Adoption of Bluetooth has been steadily increasing since its OEM automotive debut in the 2003 Saab 9-3. Although Bluetooth will continue its momentum in new 2004 models from Acura, Audi, Lexus, Lincoln, Maybach, and Toyota, to name a few, its role will be limited, according to the findings of a new research study from ABI.

Protocols based on 802.11, such as DSRC (Dedicated Short Range Communications), promise to fill the need for longer range, higher bandwidth applications that will not only link vehicles with roadside data access points, but between each other. Ratification of the new DSRC protocol is expected by the end of 2004, with aftermarket automotive offerings quickly following suit.

Aftermarket 802.11-based automotive products will appear by the end of 2003, initially focused on entertainment applications.

While the first wave of Bluetooth devices in the vehicle will centre around telephony, newer applications will soon follow. These include remote vehicle diagnostics, lower-cost telematics services, advanced automotive safety systems, vehicle-to-vehicle communications, and remote audio and video downloads into the vehicle, among others.

Its still remains to be seen if in traffic jams with multiple users using their cell phone Bluetooth hands free kits whether performance and audio degradation will be an issue.

3.5 Ethernet

Ethernet is gaining popularity in the industrial automation segment due to it being tried and tested, easy to install and maintain and less expensive than other alternatives. When first adopted in the mid-1980s as the Institute of Electrical and Electronics Engineers (IEEE) 802.3 standard, Ethernet was considered unsuitable on the plant floor. The network protocol was (and still is) nondeterministic. Device response times could not be guaranteed because of data collisions and the delays in retransmitting data. Data throughput was slow. The network medium was subject to electromagnetic interference (EMI). Ethernet now runs over shielded and unshielded twisted pair copper, coaxial, and fully EMI-resistant fibre-optic cable. Ethernet operating speeds across even conventional wire cabling have increased at least an order of magnitude, from 10 Mbps to 100 Mbps. Automatic switches that negotiate 10/100 Mbps are commonplace, thereby optimizing speed versus service, as well as letting you mix 10 Mbps (Ethernet) and 100 Mbps (Fast Ethernet) devices on the same network.

3.6 Universal Serial Bus (USB)

Universal Serial Bus (USB) has in the past been favored by aftermarket product suppliers such a Clarion (AutoPC). The transmission media is shielded twisted pair and runs up to 12 Mbps with up to 127 nodes. The future of USB use in the vehicle is uncertain as it may follow the consumer trend and be supplanted by Firewire. It must however be noted that USB can provide up to 127 whereas Firewire only provides for 16 but perhaps over 16 nodes in a vehicle for infotainment may never be exceeded.

4   Advanced Safety Systems

4.1   FlexRay

FlexRay is a new network communication system targeted specifically at the next generation of automotive applications or "by-wire" applications. The by-wire applications demand high-speed bus systems that are deterministic, fault-tolerant, and capable of supporting distributed control systems. BMW, Daimler Chrysler, Philips Semiconductors, Motorola, and the newest member, Bosch, are working together to develop and establish FlexRay as the standard for next generation applications.

The FlexRay Communication System is more than a communications protocol. It also encompasses a specifically designed high-speed transceiver, and the definition of hardware and software interfaces between various components of a FlexRay 'node'. The FlexRay protocol defines the format and function of the communication process within a networked automotive system.

The technology is designed to meet key automotive requirements like dependability, availability, flexibility, and a high data rate to complement the major in-vehicle networking standards - CAN, LIN, and MOST.

With the increasing amount of data communication between the vehicle's electronic control units (ECUs), it is important that a high data rate can be achieved. FlexRay is initially targeted for a data rate of approximately 10 Mbit/s; however, the design of the protocol allows much higher data rates to be achieved.

FlexRay, as a scalable communication system, allows synchronous and asynchronous data transmission. Depending on the application needs, the communication cycle can be purely synchronous, purely asynchronous, or a mixture of both. The synchronous data transmission enables time triggered communication to meet the requirement of dependable systems. The asynchronous transmission, based on the fundamentals of the byteflight™ protocol, allows each node to use the full bandwidth for event driven communications.

FlexRay's synchronous data transmission is deterministic with a minimum message latency and message jitter guaranteed. FlexRay supports redundancy and fault-tolerant distributed clock synchronization for a global time base, thus keeping the schedule of all network nodes within a tight, predefined precision window.

4.2 Time Triggered Protocol (TTP)

Designed for real-time distributed systems that are hard and fault tolerant, the time triggered protocol (TTP) ensures that there is no single point of failure.

TTP is the mature network solution that is low cost and can handle safety-critical applications. TTP silicon support has been available since 1998. Second generation silicon, supporting communication speeds of up to 25 Mbit/s, is available today. TTP has been adopted by aerospace companies due to its rigorous and plain safety approach. TTA Group, which is a governing body for time triggered protocol, has a membership including Audi, PSA, Renault, NEC, TTChip, Delphi & Visteon.

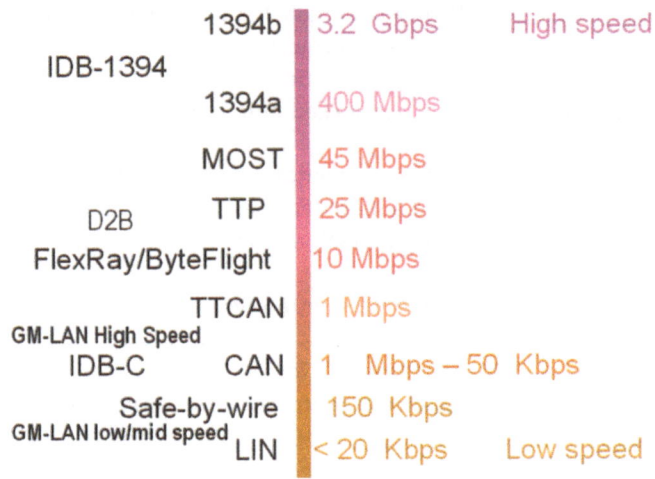

Fig. 3.   In-car network speeds.

4.3 Time Triggered Can (TTCAN)

The communication in the classic CAN network is event triggered; peak loads may occur when the transmission of several messages is requested at the same time. CAN's non-destructive arbitration mechanism guarantees the sequential transmission of all messages according to their identifier priority. For real-time systems, a scheduling analysis of the whole system has to be done to ensure that all transmission deadlines are met even at peak bus loads. To overcome this potential issue TTCAN was conceived.

TTCAN is an extension of the CAN protocol and has a session layer on top of the existing data link layer and physical layers. The protocol implements a hybrid, time triggered, TDMA schedule, which also accommodates event-triggered communications. TTCANs intended uses include engine management systems and transmission and chassis controls with scope for x-by-wire applications.

5   FPGAs Alleviate the Design Dilemma

The next few years will be a proverbial mine field for automotive electronics designers and choosing the correct data bus will be crucial to the success of integrating and testing units in production and long after the car has rolled off the assembly line. For Tier 1 suppliers and aftermarket unit design companies the problem is amplified as they will potentially supply units to many OEMs who will invariably all opt for different data busses and protocols. The industry has seen a huge shift of design philosophy away from designing a different unit for every OEM and indeed every car model to reconfigurable platforms. Reconfigurable platforms cleverly partitioned between software and reprogrammable hardware will allow the designer to change the choice of system bus or interface late in the design process and even in production. The reconfigurable system concept also enables different standards and protocols to be tried, tested and put on road trials and if you they are not found to be suitable another bus interface can be loaded into the system and tried out until the best configuration is found.

Whilst this may seem like a designers nirvana and unobtainable it can be realized today by utilizing programmable logic devices (PLDs) in the form of Field Programmable Gate Arrays (FPGA's) and Complex Programmable Logic Devices (CPLDs). PLDs can hand back to the designer the control over all phases of design from prototype, through pre-production and all phases of production. This flexibility and control can be lost when developing systems based on ASIC's and ASSP devices. PLD's can also alleviate over stocking and inventory issues as these generic devices can be used across many projects and are not application specific like ASICs. Once the programmable logic based unit is on the road it can even be reconfigured remotely via a wireless communication link to allow for system upgrades or extra functions.

The reconfigurable hardware platform can be brought to market quickly by utilizing drop in IP core blocks. For example Memec Design recently announced the availability of a cost optimised CAN core interface. The Memec CAN core contains the complete data link layer, including the framer, transmit and receive control, error core design and flexible interface enables access to each internal status and frame reference. Bit rate and sub-bit segments can be configured to meet the required timing specification of the connected CAN bus. Error counters and error interrupt events report errors. The core is designed to provide a bus bit rate of up to 1 Mbps, with a minimum core clock frequency of 8 MHz. The CAN core can provide an interface between the message filter, the message priority mechanism and various system functions such as sensor/activator control. Alternatively, it can be embedded into a system application interfacing with the microprocessor and various peripheral functions. Another example is the LIN core available from Intelliga. The LIN core, iLIN, is supplied as a reference design that uses a synchronous 8-bit general purpose micro-controller interface with minimal buffering for the transportation of message data. In addition, the reference design includes a single slave message response filter and a software interface that allows the connected micro-controller to perform address filtration.

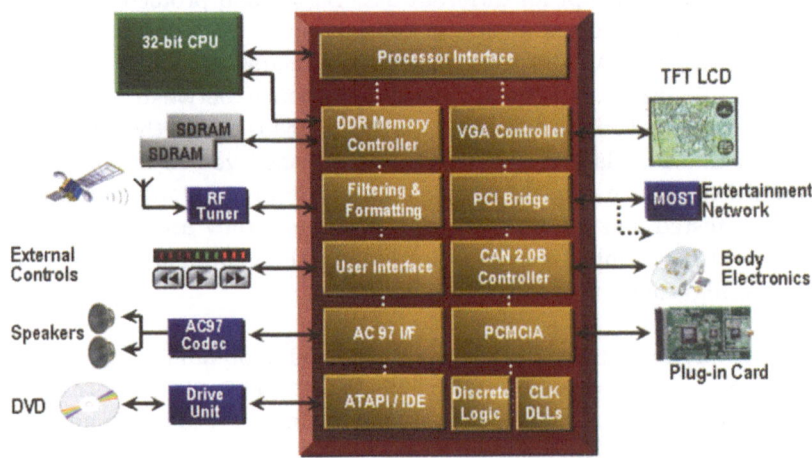

Fig. 4.   In-car multimedia design showing functions embedded in an FPGA.

IP cores can be used as part of a more complex design to provide the interface to the CAN or LIN bus instead of using a discrete device. Reducing component count has many benefits including lowering overall system cost, reducing inventory, increasing system reliability and also can help reduce PCB complexity and layers. Figure 4 shows a generic in-car multimedia design showing the use of the CAN core coupled with PCMCIA interface, PCI bridging, IDE interface plus other functions. These functions can be modified, changed or enhanced during the design phase or modified depending on the end customer requirements. In this way one PCB can be used for many customers with customisation taking place within the FPGA instead of at the board level. This model can be extended to include modification or upgrades in the field utilising a wireless connection to re-configure the FPGA in system.

6 Conclusion

It is clear that there are numerous emerging in-car bus systems providing data and control signals to and from nodes as diverse as door locks to highly sophisticated multimedia terminals. The needs of each bus are different, at the low end we need low cost relatively low speed busses and at the high end we require high speed, real time data transfer over optical media. Many automotive OEMs are backing more than one standard due to uncertainties over which one will eventually prevail. As a designer these uncertainties can delay development and ultimately lead to lost revenue. A solution designers are turning to are reconfigurable systems based on FPGAs that can be reprogrammed to accommodate changing standards and protocols late in the design process and even in production. The bus interface can take the form of pre-verified drop in IP Cores that can safe time and effort and thus increase time to market.

References

[1] Costlow, T., Automotive Networking, Automotive Engineering International, May 2003.

[2] Gould, L., Industrial Ethernet: Wiring the Enterprise, Automotive Design & Production, August 2003.

[3] Hefferman, D. and Leen, G., Expanding Automotive Electronic Systems, IEEE 2002.

[4] Lupini, C, A., Multiplex Bus Progression, SAE Technical Paper Series, January 2001.

[5] Whitfield, K., Broadband for the Car, Automotive Design & Production, June 2003.

Website References

CAN	www.can.bosch.com
CAN IP	www.memecdesign.com/can_core
GM LAN	www.gmtcny.com/lan.htm
MOST	www.mostnet.de
TTTech (TTP)	www.ttagroup.org, www.ttchip.com
FlexRay	www.flexray.com
D2B	www.candc.co.uk
LIN	www.lin-subbus.org
LIN IP Core	www.intelliga.co.uk
Bluetooth	www.xilinx.com/esp/technologies/wireless_networks/bluetooth.htm
IDB	www.idbforum.org
Ref. Boards	www.xilinx.com/xlnx/xebiz/board_search.jsp

Karen Parnell
Xilinx
203 Brooklands Road
Weybridge
Surrey
United Kingdom
Karen.parnell@xilinx.com

Keywords: Reconfigurable hardware, upgradeable telematics, in vehicle standards and protocols

Appendix A

List of Contributors

List of Contributors

Appendix B

List of Keywords

List of Keywords